Graphing Calculator
for
NYS Math A ...
and Beyond

By
Kathleen Noftsier

Note for Librarians: A cataloguing record for this book is available from Library and Archives Canada at www.collectionscanada.ca/amicus/index-e.html

ISBN 1-4251-0138-0

Trafford's print shop runs on "green energy" from solar, wind and other environmentally-friendly power sources.

TRAFFORD
PUBLISHING™

Offices in Canada, USA, Ireland and UK

Book sales for North America and international:
Trafford Publishing, 6E–2333 Government St.,
Victoria, BC V8T 4P4 CANADA
phone 250 383 6864 (toll-free 1 888 232 4444)
fax 250 383 6804; email to orders@trafford.com

Book sales in Europe:
Trafford Publishing (UK) Limited, 9 Park End Street, 2nd Floor
Oxford, UK OX1 1HH UNITED KINGDOM
phone +44 (0)1865 722 113 (local rate 0845 230 9601)
facsimile +44 (0)1865 722 868; info.uk@trafford.com
Order online at:
trafford.com/06-1895

10 9 8 7 6 5 4 3 2 1

Table of Contents

Graphing Calculator for Math A and Beyond

This graphing calculator workbook is designed to:
1. Give you alternate strategies for solving Math A and B questions.
2. Give extra practice on regents-type questions.
3. Demonstrate how to work around "quirks" in the programming of the calculator.
4. Begin developing skills needed for the Math B exam, and college calculus and statistics courses.

**Note that some topics go beyond the scope of the Math A exam. This is to help justify giving math credit for those schools that choose to use this as a text for a mathematics elective. They can be omitted if the workbook is used as a supplement to a course leading to the Math A exam.

The exercises in this book were designed to be completed with the **TI-83+/TI-84+** calculator. Most will work with the TI-83 or TI-82. Advantages to the TI-83+/TI-84+ include ability to upgrade the operating system and download applications from the TI website or other calculators.

Beware!! These calculators are designed to work with TI applications only. Do not download non-TI applications or programs to your calculator. It can cause your calculator to work inefficiently or disable your operating system.

Protect your calculator! Whether you have your own, or are responsible for a school-owned calculator, you should know how to identify the calculator.

The TI-83+/TI-84+ have two identification numbers.
1. The first is on the outside. Find the number on the back of the calculator you will be responsible for (yours or the school's) and record it here: _____

2. The second is an internal number. To find it:
   a. Turn the calculator on.
   b. Go to Memory (2nd, +).
   c. Choose **1:About**
   d. Look for **ID:**
   e. Write the calculator's ID# here: _____

f. Your signature: _____
g. This is also the number you will need to identify your
    calculator for downloads from the TI website.

You should also note the number under the product name. This is the
number of the current operating system (OS). The operating system
should be updated so that the most current system is loaded to
ensure the best performance.

At this time, the most current OS is: _____

If your calculator has an OS lower than this it should be updated at the
earliest opportunity.

A note about the pages marked

These are collections of all Math A regents questions (and some Math
B questions) to date (12/03) on the topic immediately preceding the
section. **They are not all intended to be done on the graphing
calculator**. They are meant as an opportunity to practice questions
for the regents exam. They should also provide an opportunity to
discern which questions can be done on the calculator for the Math A
exam and which should be completed using other methods.

Good luck and happy calculating!

Basic Operations

The graphing calculator will work much like a scientific calculator for most of the familiar calculations. There are some things to watch for though.

1. Note that for some operations the calculator will insert parentheses with the operation. Try the square roots below. Allow the calculator to insert the beginning parentheses; you do not need to repeat them. Just insert the ending ones where they are printed.

   a. $\sqrt{(4} +3$ _____   b. $\sqrt{(4)} +3$ _____

   Other operations that will work in a similar way include the trigonometry functions.

2. We must also be careful with negative numbers.
   a. Be sure to use the negative key found below the number 3 and not the subtraction key. If the calculator says "Syntax Error" check to see if you have used the incorrect one in your calculation. Most syntax errors are either this mistake or mismatched parentheses.
   b. Write the name for the following number as you would say it:

   $$-5$$

   _____

   The calculator does not read the number this way. When the calculator sees –5, it reads –1 multiplied by 5, (-1 • 5). In most cases this will not make a difference, but try the calculations below as they are written. (Use the x^2 key.)

   $-3^2 =$_____ $(-3)^2 =$_____

   Use the explanation above and your knowledge of the order of operations for real numbers to explain why the answers are different.

   _____

   _____

3. If you need to raise a number or variable to a power other than 2 you can use the power key found between the ÷ key and the CLEAR key. It works similarly to the x^2 key.

Practice on these: (Use parentheses only when they are printed.)

a. $2^5=$ _____   b. $-2^5=$ _____   c. $(-2)^5$ _____

d. $2^4=$ _____   e. $-2^4=$ _____   f. $(-2)^4$ _____

g. $2^3=$ _____   h. $-2^3=$ _____   i. $(-2)^3$ _____

j. $3^6=$ _____   k. $-3^6=$ _____   l. $(-3)^6$ _____

Explain why the last column sometimes matches the first column and other times matches the middle column.

_____

_____

_____

4. Two last very useful items:
   a. Pressing 2nd and the negative key will **bring back the answer to the previous calculation** at the cursor's current location. This is especially nice to have if you need to use it in a calculation but it is a repeating decimal or an irrational number. Your calculations will be more accurate if you use this unrounded answer when you can.

   b. Pressing 2nd and ENTER will **"back-space"** for you. Repeat until you reach the calculation you wish to use again. Unlike the 2nd answer, 2nd entry will continue back as far as its short-term memory allows. It could save more than 20 simple calculations or as few as 2 or 3 complicated calculations.

   As the Regents exam requires you to wait until your final calculation to round, these could be especially helpful in retaining long, unrounded decimals for use in a complicated calculation. Another tool for this will be to store values to variables, but we'll save that for a later lesson.

"Where's the fraction key?"

"Sorry – there isn't one."

Wait!! Don't reach for that scientific calculator yet. We can easily work with fractions on the TI-83+/TI-84+ without a special key.

Remember – a fraction is a _____ problem.
The division key works nicely to make a simple fraction.

Example: $\frac{7}{12}$ is entered as | 7 | | ÷ | | 1 | | 2 |

You will get a decimal answer.

Basic Operations with Simple Fractions: $\frac{2}{3} + \frac{4}{5}$

Enter as 2/3+4/5 where / is entered on the screen with the division key.

If you need your answer in fraction form press | MATH |,

choose 1: ►Frac, press | ENTER | on the home screen.

In the example above, the result is an improper fraction. **If you need to change this fraction to a mixed number:**

1. Note the number in front of the decimal before changing to a fraction.
2. Write this as the whole number part of your mixed number.
3. Then subtract this number from the decimal.
4. Convert the decimal that is left to a fraction.
5. Use this fraction as the fractional part of the mixed number.

Mixed Numbers:

Mixed numbers can be treated as grouped addition problems.

$5\dfrac{1}{6}$ is 5 whole parts plus 1/6.

We will enter this in the calculator as (5+1/6).

***Note the parentheses. They are needed to keep the parts of the mixed number grouped.

***Also note that if the mixed number is negative, the negative symbol MUST come before (outside) the parentheses!!

Try each of the following problems with and without parentheses.

a. $2\dfrac{1}{3}+4\dfrac{5}{7}$ b. $2\dfrac{1}{3}-4\dfrac{5}{7}$ c. $2\dfrac{1}{3}\cdot4\dfrac{5}{7}$ d. $2\dfrac{1}{3}\div4\dfrac{5}{7}$

Answer with
Parentheses:

a. _____   b. _____   c. _____   d. _____

Answer without
Parentheses:

a. _____   b. _____   c. _____   d. _____

What conclusion can be made based on your results?

Why were the results different based on the operation?

Algebraic fractions (fractions with variables in) must also be treated carefully so that the operations are done in the correct order and a variable stays grouped with its coefficient. (A coefficient is

_____).

Example: Pick a number larger than 1 and store it to x.
(Suppose you chose 5, 5 STO▶ X.)

If you begin with the fraction $\frac{3}{8x}$, try entering this fraction as 3/8x and

as 3/(8x). Are they the same? _____

Which one do you think is correct? _____

Remember that the fraction bar is a grouping symbol. You need to group the numerator because **the calculator only recognizes division between the character immediately before and after the symbol**. All other characters are treated with the normal order of operations.

Practice:

Convert to a decimal:

1. $\frac{7}{8}$ _____   2. $\frac{4}{3}$ _____   3. $\frac{2}{9}$ _____

Use ┌─────────┐
    │ MATH   │   1: Frac to convert the following decimals to fractions.
    └─────────┘

4. .625 _____   5. .44 _____   6. .52 _____

2 2/3 +1/5

Express answers as both improper fractions and mixed numbers when possible.

Add.

7. $\dfrac{1}{3}+\dfrac{4}{5}$ _____

8. $\dfrac{2}{5}+\dfrac{7}{8}$ _____

9. $\dfrac{1}{7}+\dfrac{1}{11}$ _____

10. $\dfrac{4}{9}+\dfrac{6}{13}$ _____

11. $-2\dfrac{2}{5}+3\dfrac{1}{9}$ _____

12. $1\dfrac{3}{4}+4\dfrac{6}{7}$ _____

13. $4\dfrac{5}{8}+6\dfrac{1}{6}$ _____

14. $-12\dfrac{3}{17}+21\dfrac{1}{12}$ _____

$1/6 \sim 4\ 5/7$

Subtract.

15. $\dfrac{3}{4}-\dfrac{1}{8}$ _____

16. $\dfrac{2}{5}-\dfrac{1}{6}$ _____

17. $\dfrac{1}{9}-\dfrac{2}{3}$ _____

18. $\dfrac{2}{5}-\dfrac{5}{11}$ _____

19. $-2\dfrac{3}{5}-1\dfrac{5}{7}$ _____

20. $5\dfrac{1}{10}-\dfrac{7}{8}$ _____

21. $1\dfrac{1}{6}-3\dfrac{4}{13}$ _____

22. $-12\dfrac{3}{5}-31\dfrac{5}{16}$ _____

Multiply.

23. $\dfrac{1}{6} * \dfrac{4}{5}$ _____

24. $\dfrac{2}{7} * \dfrac{4}{11}$ _____

25. $\dfrac{11}{13} * \dfrac{2}{19}$ _____

26. $\dfrac{3}{20} * \dfrac{4}{7}$ _____

27. $4\dfrac{1}{8} * 5\dfrac{2}{9}$ _____

28. $-7\dfrac{1}{3} * 10\dfrac{6}{7}$ _____

29. $11\dfrac{4}{9} * 4\dfrac{11}{15}$ _____

30. $20\dfrac{3}{8} * 5$ _____

Divide.

31. $\dfrac{3}{4} \div \dfrac{1}{7}$ _____

32. $\dfrac{5}{8} \div \dfrac{2}{3}$ _____

33. $\dfrac{1}{4} \div \dfrac{2}{7}$ _____

34. $\dfrac{4}{9} \div 7$ _____

35. $-2\dfrac{6}{7} \div \dfrac{1}{9}$ _____

36. $4\dfrac{2}{3} \div 1\dfrac{1}{11}$ _____

37. $4 \div 6\dfrac{3}{8}$ _____

38. $9\dfrac{5}{17} \div 15\dfrac{2}{5}$ _____

Challenge:

(Note: If one of these will not change to a mixed number for you, write the entire decimal you see on the screen.)

39. $\dfrac{1}{3} + 4\dfrac{4}{9} * 5\dfrac{1}{6} - 1\dfrac{2}{7} \div \dfrac{4}{13}$ _____

40. $2\dfrac{11}{23} - 1\dfrac{11}{19} * 33\dfrac{1}{18} + 14\dfrac{33}{34} \div 112\dfrac{1}{8}$ _____

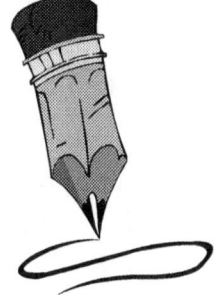

 ## Regents Connection 1

Basic Operations

<u>June '99, #11:</u> The expression $2^3 \cdot 4^2$ is equivalent to

    (1) 2^7 (2) 2^{12} (3) 8^5 (4) 8^6

<u>June '99, #20:</u> The expression $\sqrt{27} + \sqrt{12}$ is equivalent to

    (1) $5\sqrt{3}$ (2) $13\sqrt{3}$ (3) $5\sqrt{6}$ (4) $\sqrt{39}$

<u>Jan '00, #1:</u> The expression $\sqrt{93}$ is a number between

    (1) 3 and 9   (3) 9 and 10
    (2) 8 and 9   (4) 46 and 47

<u>Jan '00, #2:</u> Which number has the greatest value?

    (1) $1\frac{2}{3}$ (2) $\sqrt{2}$ (3) $\frac{\pi}{2}$ (4) 1.5

<u>Aug '99, #2:</u> the expression $\sqrt{50}$ can be simplified to

    (1) $5\sqrt{2}$ (2) $5\sqrt{10}$ (3) $2\sqrt{25}$ (4) $25\sqrt{2}$

<u>Aug '99, #8:</u> The formula $C=\dfrac{5}{9}(F\text{-}32)$ can be used to find the Celsius temperature (C) for a given Fahrenheit temperature (F). What Celsius temperature is equal to a Fahrenheit temperature of 77°?

(1) 8°   (2) 25°   (3) 45°   (4) 171°

<u>Aug '99, #14:</u> In a hockey league, 87 players play on seven different teams. Each team has at least 12 players. What is the largest possible number of players on any one team?

(1) 13   (2) 14   (3) 15   (4) 21

<u>June '00, #20:</u> What is the value of 3^{-2}?

(1) $\dfrac{1}{9}$ (2) $-\dfrac{1}{9}$ (3) 9   (4) –9

<u>June '00, #21:</u> the formula for changing Celsius (C) temperature to Fahrenheit (F) temperature is $F=\dfrac{9}{5}C+32$. Calculate, to the nearest degree, the Fahrenheit temperature when the Celsius temperature is – 8.

<u>Sample #19:</u> The expression $\sqrt{150}$ is equivalent to

(1) $25\sqrt{6}$ (2) $15\sqrt{10}$ (3) $5\sqrt{6}$ (4) $6\sqrt{5}$

<u>Jan '03, #11:</u> The sum of $\sqrt{75}$ and $\sqrt{3}$ is

(1) 15   (2) 18   (3) $6\sqrt{3}$ (4) $\sqrt{78}$

June '03, #12: The expression $3^2 \cdot 3^3 \cdot 3^4$ is equivalent to

(1) 27^9 (2) 27^{24} (3) 3^9 (4) 3^{24}

June '03, #16: The sum of $\sqrt{18}$ and $\sqrt{72}$ is

(1) $\sqrt{90}$ (2) $9\sqrt{2}$ (3) $3\sqrt{10}$ (4) $6\sqrt{3}$

Benefits: Resetting the calculator erases any changes a previous user may have made which can affect your results.

Disadvantage: Properly clearing the memory will delete all unarchived programs and most applications.

You may not want your own calculator reset!!!

To reset:

1. | 2nd | | + | (Memory)

2. Choose 7:Reset... (5:Reset on a TI83)
3. The following screen will appear:

a. To eliminate changes to basic operations choose **2:Defaults.**

b. To eliminate changes to basic operations AND any programs choose **1:All RAM.**

c. To clear only the archive use the right arrow key to move the cursor to this choice.

d. To clear **everything** use the right arrow key and choose **ALL.** ENTER.

The calculator should be completely reset including programs and archive before the Regents exam!

Therefore, we will choose ALL.

```
RAM ARCHIVE ALL
1:All Memory...
```

Choose 2:Reset.

Correct:

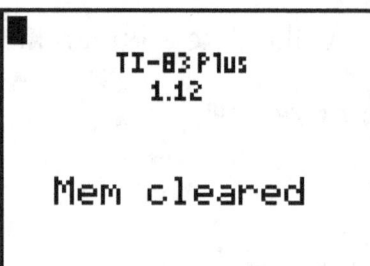

Incorrect:

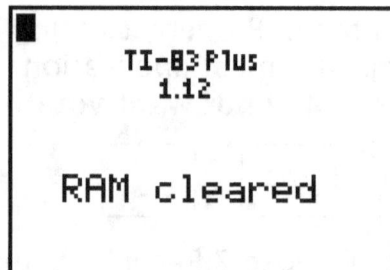

When taking the Regents you may want to:

1. Back up your personal programs and applications on another calculator or a computer

Or

2. Use a calculator belonging to the school for the exam if your school is able to provide them.

Caution! If your school uses the TI Navigator on a regular basis you may wish to reset only the RAM as a RESET ALL will delete the Applications necessary to connect to the Navigator.

Intro to Equation Solver:

Solver

Equation Solver can be used to find one unknown in any equation. (Remember, to be an equation it must have _____ .)

To use Solver with a TI-83+/TI-84+ graphing calculator:

1. Press the [MATH] key.

2. Choose 0:Solver... from the MATH menu.

3. "eqn: 0=" should appear. If there is an equation already entered press the up arrow then [CLEAR]

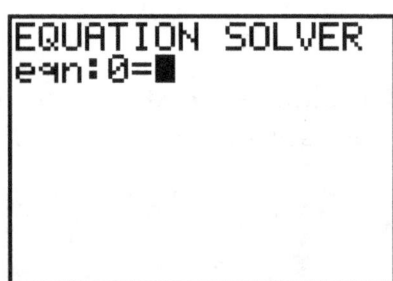

   Be careful not to press CLEAR on the variable. If the variable has no value entered the calculator will not allow you to "move up" to the equation to change it. Enter any number in the variable and it will allow you to continue.

4. Key in the equation you wish to solve, but use a subtraction symbol in place of the equal sign and, if there is more than one term after the equal sign, enclose anything on the right side of the equal sign in parentheses. This sets your equation equal to zero.

5. Press ENTER .

6. If there is only one variable in your equation go on to #7.

    If there are two or more variables you must enter the values for all but the one you are solving for.

7. Press ALPHA .

8. Press ENTER . (Solve)

9. If the unknown is your answer you are finished. If you need to insert it into another equation continue.

10. Press 2nd .

11. Press MODE . (Quit)

12. On the home screen enter the expression or equation. The calculator will remember what the value of the variable was so you do not need to enter it, **use the variable itself where it appears in the expression**.

Caution: Be sure to follow each step! Previous values for the variables remain until you have "solved".

ERROR!

Error messages on the TI-83+/TI-84+ are aggravating, but they are usually _easy to correct_.

These are the most frequently seen error messages when using SOLVER:

> ERR:NO SIGN CHNG
> ERR:DIVIDE BY 0
> ERR:SYNTAX
> ERR:DATA TYPE
> ERR:ITERATIONS

SOLVER works by starting with the current value of the variable being solved for and uses a "guess-and-check" type of system to work its way closer and closer to the correct value until it has solved the equation.

The first two error types,

> ERR: NO SIGN CHNG & ERR:DIVIDE BY 0

occur frequently because we do not notice that the denominator of a fraction may start or pass through a value of zero during the solving process. The calculator recognizes that it is not possible to divide by zero, tries to work around the problem, then gives up.
When the error message appears it also gives two choices:

> 1:QUIT
> 2:GOTO

Choosing quit will erase the incorrect equation completely. **_Don't do this!!_**

Choosing GOTO (Go to) will take you to the place where the calculator believes the error is occurring. With the first two errors, **_changing the beginning value of your variable should correct the error._**

Example:

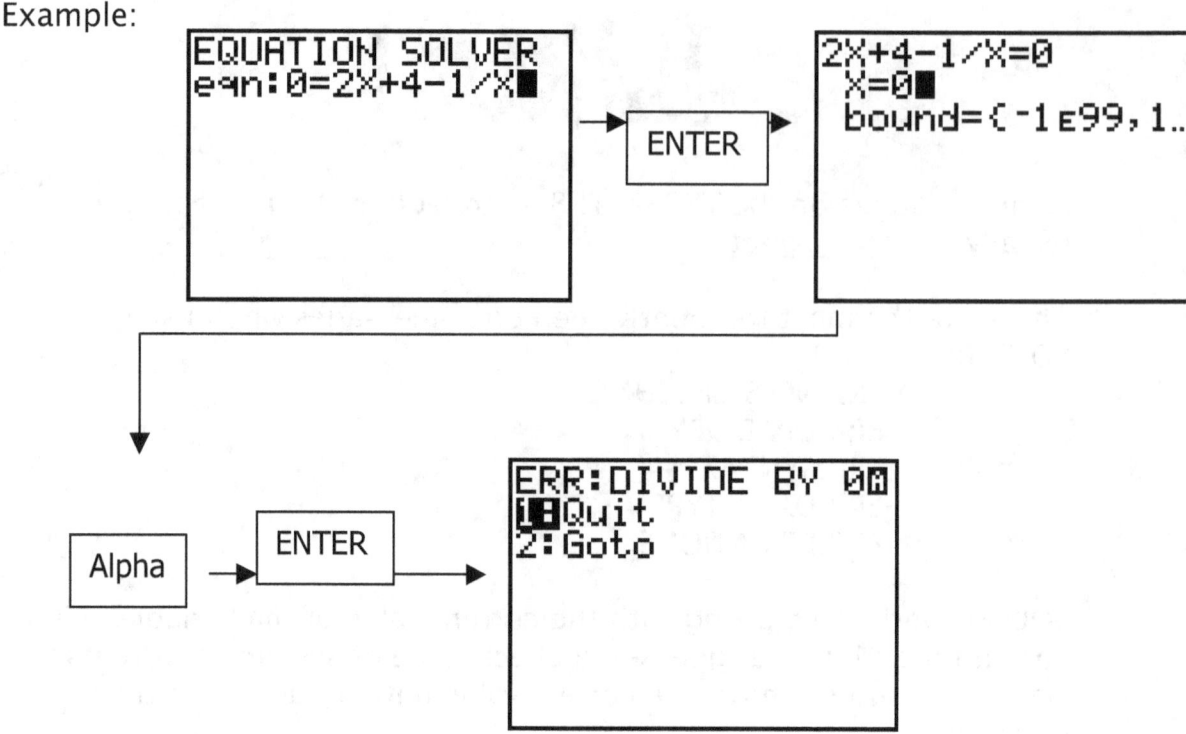

Choosing "2:GOTO" will take you back to the second screen and let you change the beginning X= to a value other than 0.

A "SYNTAX" error usually means that you are missing parentheses. In particular, you probably have an ending parenthesis, ")", with no matching beginning one,"(".

Or you may have used a subtraction symbol in place of a negative sign where the calculator cannot adjust for this mistake.

The GOTO command will take you to the place where the calculator thinks the error is, but you should check through the equation yourself to see if you need to add something or delete an extra character.

A DATA TYPE error occurs if you are trying to input something that won't work in that program. For example, you cannot enter a program name into solver; it is not an equation.

An ITERATIONS error occurs if the beginning value for the variable is too far from the actual value. The calculator is programmed to solve within a certain number of "guess-and-checks" (iterations). If it cannot find the answer within that number of tries it gives up.

Try another value for the variable. If this does not work try graphing the equation to see if the solution exists or is very large or small (negative). Seeing the graph may help you to choose a more appropriate beginning value (first guess) for the variable.

Some less frequently seen errors:
1. **ERR:SINGULARITY**. Although it is a common error, it is unusual to receive this message. The calculator will usually treat it as a NO SIGN CHANGE error. The equation in undefined at some point.

2. It is also possible to get an **ERR:NONREAL ANS** in Equation Solver. Equation Solver will not work with complex (imaginary) numbers. [This will make more sense when you take Math B!]

3. Finally, if the equation is undefined at the initial value of x and several points around this value, you may see
ERR:BAD GUESS.

If you notice that the calculator seems to be taking a long time to solve an equation, you can tell it is still working by the "dots" moving in the upper right corner of the screen (an almost certain sign that an error message is on its way), and you think you know what the problem is, you can interrupt the calculator by pressing the `ON` **key.**

Solving for Zero

Because the equation solver on the graphing calculator will always start "0=...", every equation must be set equal to zero before beginning.

Example:
$$2x+3=9$$
$$\underline{-9\quad-9}$$
$$2x-6=0 \qquad \text{Then } 2x-6 \text{ would be entered in solver.}$$

Find the solutions to the following equations with pencil and paper.

$$3x + 8 - 10 + 2x = 0 \qquad\qquad 3x + 8 - (10 + 2x) = 0$$

Are they equal? _____   Why or why not?

_____

_____

We will use a "short-cut" to set our equations equal to zero. The demonstration above should point out the importance of parentheses in the "short-cut" method.

Set $6x + 7 = 3x - 4$ equal to zero using the method in the example.

Is this the same as $6x + 7 - (3x - 4) = 0$? _____

Is this the same as $0 = 6x + 7 - (3x - 4)$? _____

What properties of the real numbers are we using?

_____

_____

To solve $6x + 7 = 3x - 4$ in solver, replace the "=" with a "-" and place the right side of the equation in parentheses.

Grouping Symbols

In our "order of operations", which operation takes precedence over all others?

Although we usually only name parentheses, any type of grouping must take precedence (be taken care of before) any other operation.

Name some other types of grouping symbols:

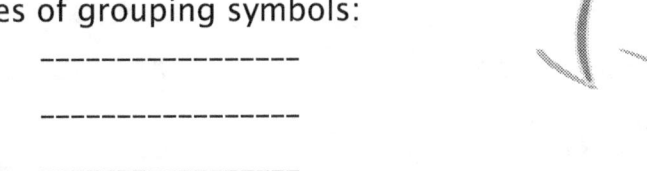

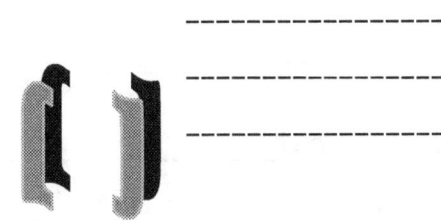

The graphing calculator sometimes needs help recognizing those things we assume are grouped.

Example 1:   Consider the equation $\dfrac{1}{2x} = 20$

Entered as it is in the calculator it would appear as 1/2x=20.
*****The calculator does not read this as
                "1 divided by 2x is 20."
It is read as
                "½ times x is 20."

This should be entered as 1/(2x)=20.

Example 2:   Let's look at $2x + 3 = 4 - \dfrac{2}{3}x$

When we enter this equation in Solver we need extra parentheses. Try
2x+3-(4-(2/3)x)   x=_____

Why do we need the parentheses before 4 and after x?

_____

Solver Practice

Solve for each variable below (you may use the letter given or use "x" in Solver).

a. $-6 - m = 18$ m=_____

b. $-3m = 6 - 5m$ m=_____

c. $7x + 8 = 5x + 17 - x$ x=_____

d. $3x - (6 - x) = 18$ x=_____

e. $2(x - 3) = 1.2 - x$ x=_____

f. $1.4q - 2 = 1.2q + 0.004$ q=_____

g. $2/3a - 5/6 = 1/2a$ a=_____

h. $m\angle A = 3x + 40$
   $m\angle B = x + 5$
   $m\angle C = 2x - 15$

   Find the measures of the <u>three</u> angles.

   A: _____   B: _____   C: _____

i. $m\angle 1 = x + 20$
   $m\angle 2 = 2x + 50$
   $m\angle MRT = 160$
   <u>Find $m\angle 1$ and $m\angle 2$.</u>

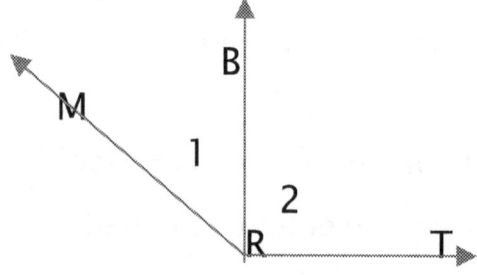

   1: _____   2: _____

Review:

1. Use SOLVER to solve the following equations:
   a. $2x + 12 = 96$ x=_____

   b. $3x - 5 = 2x + 30$ x=_____

2. a. Subtract: $2\,1/3 - 5/6$ _____

   b. Using the calculator, change the answer of part (a)
      to a fraction.

                                                              _____

3. Check your answers with a neighbor. If you do not   agree,
   compare work and decide whose work is correct. Then make
   any necessary changes.

More SOLVER:

A. Solve using pencil and paper: (Show all work!)

$$\frac{x+2}{5} = \frac{2x-30}{15}$$

          Now solve with SOLVER   x=_____
*** Proportions work best if you set up the cross multiplication
first!! This will eliminate problems caused by values of x at which
the ratio is undefined.

Enter the above example as 15(x+2)-5(2x-30). There is no need to
distribute or combine like terms. As long as you have grouped
correctly, the calculator will take care of the rest.

For B through I, use SOLVER to find the unknown.
If the solution is a decimal, use the calculator to **change the answer to a proper fraction or mixed number**. For best results, set up the cross multiplication first, then solve.

B. $\dfrac{x-6}{3x+18} = \dfrac{2}{5}$ x=_____

C. $\dfrac{x+1}{3x-12} = \dfrac{2}{7}$ x=_____

D. $\dfrac{5x-1}{3} = \dfrac{2x+13}{4}$ x=_____

E. $\dfrac{11x-3}{4} = \dfrac{x+5}{19}$ x=_____

F. $\dfrac{x+2}{27} = \dfrac{x-3}{2}$ x=_____

G. $\dfrac{4x+5}{11} = \dfrac{x-3}{2}$ x=_____

H. $\dfrac{3}{x-5} = \dfrac{13}{x}$ x=_____

I. $\dfrac{2}{x+1} = \dfrac{15}{3x}$ x=_____

SOLVER SOLVER SOLVER

Decimal Solutions to Fractions

To change a decimal SOLVER answer to a fraction:

1. Enter equation and solve.

2. | 2nd | → | MODE | (Quit)

3. If the variable solved for was X, | X,T,θ,n |

   Otherwise enter the variable name.

4. | MATH | → 1: Frac ▸ | ENTER | → | ENTER |

5. To change an improper fraction to a mixed number:
   - a.) Numerator ÷ Denominator
   - b.) Write down the whole number part of the decimal.
   - c.) Subtract this number.
   - d.) Change the resulting decimal to a fraction for the fractional part.

Example: $^{38}\!/_8$

1. 38 | ÷ | 8

2. | ENTER | — | 4 | ENTER |

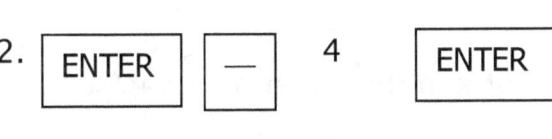

3. | MATH | 1:▸ Frac

4. | ENTER |

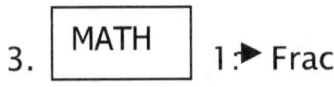

Therefore, $\dfrac{38}{8} = 4\dfrac{3}{4}$.

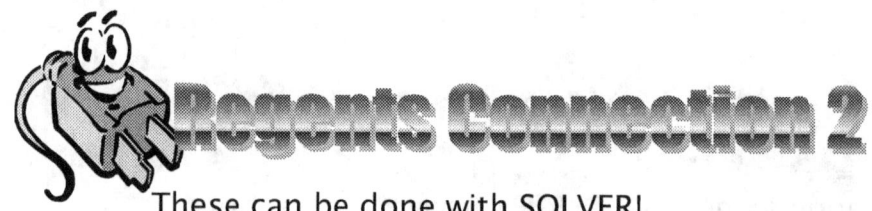

These can be done with SOLVER!

Sample '98, #2: If 12x=4(x+5), then x equals

    (1) $\dfrac{1}{12}$ (3) 1.25

    (2) $\dfrac{5}{8}$ (4) 2.5

Aug '99, #21: Solve for x: 2(x-3)=1.2-x

Aug '00, #15: Solve for x: 15x-3(3x+4)=6

    (1) 1   (3) 3

    (2) $-\dfrac{1}{2}$ (4) $\dfrac{1}{3}$

Jan '02, #10: There are 357 seniors in Harris High School. The ratio of boys to girls is 7:10. How many boys are in the senior class?

    (1) 210   (3) 117

    (2) 147   (4) 107

Jan '02, #4: What is the value of x in the equation $\dfrac{3}{4}x+2=\dfrac{5}{4}x-6$?

    (1) –16   (3) –4

    (2) 16   (4) 4

Jan '02 #22: A 12-foot tree casts a 16-foot shadow. How many feet tall is a nearby tree that casts a 20-foot shadow at the same time?

<u>Jan '02, #29:</u> In the accompanying diagram, $\overleftrightarrow{AB}$ and $\overleftrightarrow{CD}$ intersect at E. If m∠AEC=4x-40 and m∠BED=x+50, find the number of degrees in∠AEC.

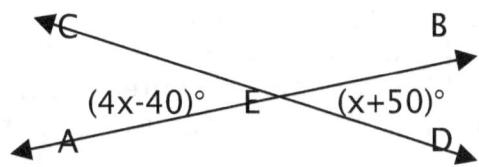

<u>June '01, #11:</u> If one-half of a number is 8 less than two-thirds of the number, what is the number?

    (1) 24   (2) 32   (3) 48   (4) 54

<u>Jan '01, #28:</u> In the accompanying figure, two lines intersect, m∠3=6t+30, and m∠2=8t-60. Find the number of degrees in m∠4.

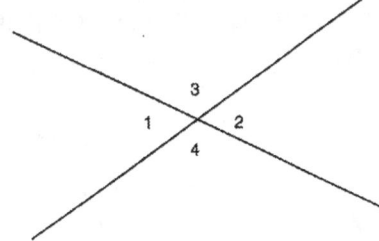

<u>Aug '00, #19:</u> A girl can ski down a hill five times as fast as she can climb up the same hill. If she can climb up the hill and ski down in a total of 9 minutes, how many minutes does it take her to climb up the hill?

<u>Aug '00, #24:</u> The sum of the ages of the three Romano brothers is 63. If their ages can be represented as consecutive integers, what is the age of the middle brother?

<u>June '02, #1:</u> Jamie is 5 years older than her sister Amy. If the sum of their ages is 19, how old is Jamie?

<u>June '02, #14:</u> What is the solution of the equation 3y-5y+10=36?

   (1) –13   (2) 2   (3) 4.5   (4) 13

<u>June '02, #26:</u> Two parallel roads, Elm Street and Oak Street, are crossed by a third, Walnut Street, as shown in the accompanying diagram. Find the number of degrees in the acute angle formed by the intersection of Walnut Street and Elm Street.

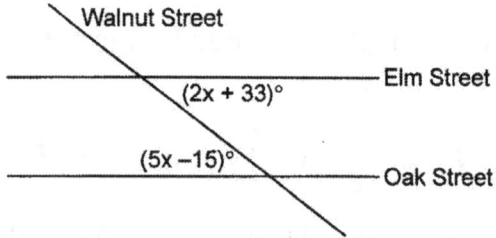

<u>Aug '02, #2:</u> In the accompanying diagram of parallelogram ABCD, diagonals $\overline{AC}$ and $\overline{DB}$ intersect at E, AE=3x-4, and EC=x+12.

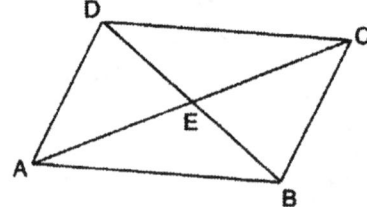

What is the value of x?

   (1) 8   (2) 16   (3) 20   (4) 40

<u>Aug '02, #19:</u> If 2x+5=-25 and –3m-6=48, what is the product of x and m?

   (1) –270   (2) –33   (3) 3   (4) 270

<u>Sample #15:</u> In the diagram below, $\overline{AB}$ is parallel to $\overline{CD}$. Transversal $\overline{EF}$ intersects $\overline{AB}$ and $\overline{CD}$ at G and H, respectively. If m∠AGH=4x and m∠GHD=3x+40, what is the value of x?

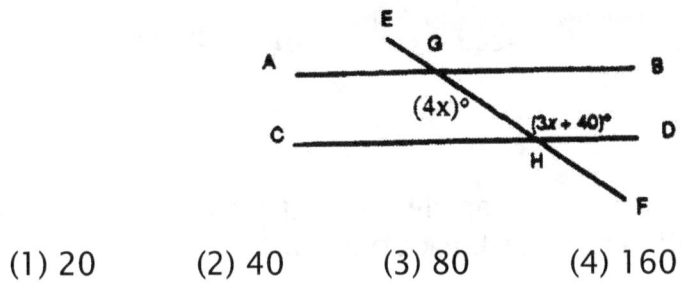

(1) 20   (2) 40   (3) 80   (4) 160

<u>June '03, #23:</u> Solve for m: 0.6m+3=2m+0.2

<u>June '03, #24:</u> In the accompanying diagram, line m is parallel to line p, line t is a transversal, m∠a=3x+12, and m∠b=2x+13. Find the value of x.

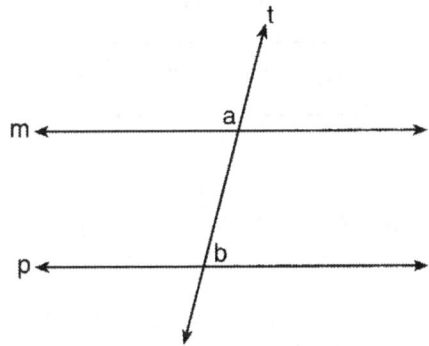

in SOLVER

Although percents may be more easily done "the old-fashioned-way", SOLVER can be a useful check or back-up method.

So far we have used only "x" as our variable. SOLVER will allow us to use several variables at once in an equation. This is very useful in formulas.

Our first formula is a familiar one:

$$\frac{IS}{OF} = \frac{\%}{100}$$

We can only use one letter for our unknown so, to keep things simple, we will use the first letter of each word:

I=is   (| ALPHA | | X^2 |)

O=of   (| ALPHA | | 7 |)

P=%   (| ALPHA | | 8 |)

In SOLVER, enter (I/O)-(P/100) after 0=   .

The next screen will no longer list "x=". It should now say:

I=
O=
P=

SOLVER can only solve for one variable at a time. We need to consider a specific example:

A. What % of 64 is 4?

$$I = is = _____$$
$$O = of = _____$$
$$P = \% = _____$$

    1. Enter the two variables you can identify from the question.
    2. Using the up and down arrows align the cursor with the third variable (in this case P).
    3. SOLVE.

Complete the following using the same method:

B. What % of 125 is 5?

$$I=_____$$
$$O=_____$$
$$P=_____$$

C. 6 is 30% of what number?

$$I=_____$$
$$O=_____$$
$$P=_____$$

D. 2 is 1% of what number?

$$I=_____$$
$$O=_____$$
$$P=_____$$

E. What is 11% of 45?

$$I=_____$$
$$O=_____$$
$$P=_____$$

F. What is 107% of 23?

$$I=_____$$
$$O=_____$$
$$P=_____$$

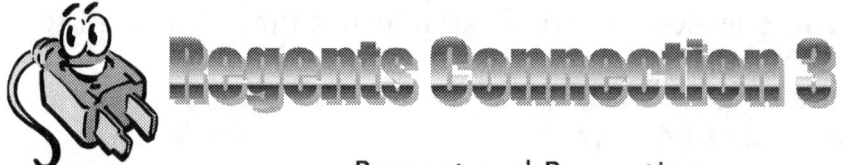

Percent and Proportion

<u>June '00, #24:</u> The Rivera family bought a new tent for camping. Their old tent had equal sides of 10 feet and a floor width of 15 feet, as shown in the accompanying diagram.

Old Tent

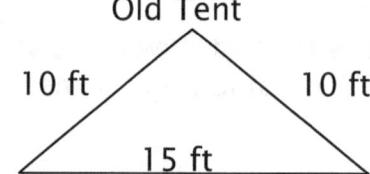

10 ft   10 ft

15 ft

If the new tent is similar in shape to the old tent and has equal sides of 16 feet, how wide is the floor of the new tent?

<u>Aug '99, #15:</u> In the accompanying diagram of equilateral triangle ABC, DE=5 and $\overline{DE} \parallel \overline{AB}$.

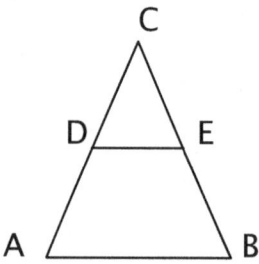

If AB is three times as long as DE, what is the perimeter of quadrilateral ABED?

(1) 20   (3) 35
(2) 30   (4) 40

<u>Aug '99, #31:</u> The profits in a business are to be shared by the three partners in the ratio of 3 to 2 to 5. The profit for the year was $176,500. Determine the number of dollars each partner is to receive.

<u>Jan '01, #22:</u> Sue bought a picnic table on sale for 50% off the original price. The store charged her 10% tax and her final cost was $22.00. What was the original price of the picnic table?

<u>Jan '00, #9:</u> Twenty-five percent of 88 is the same as what percent of 22?

(1) 12 ½%		(3) 50%
(2) 40%		(4) 100%

<u>Jan '00, #14:</u> Sterling silver is made of an alloy of silver and copper in the ratio of 37:3. If the mass of a sterling silver ingot is 600 grams, how much silver does it contain?

(1) 48.65 g		(3) 450 g
(2) 200 g		(4) 555 g

<u>Aug '99, #30:</u> A painting that regularly sells for a price of $55 is on sale for 20% off. The sales tax on the painting is 7%. Will the final total cost of the painting differ depending on whether the salesperson deducts the discount before adding the sales tax or takes the discount after computing the sum of the original price and the sales tax on $55?

<u>June '99, #10:</u> Linda paid $48 for a jacket that was on sale for 25% of the original price. What was the original price of the jacket?

<u>June '99, #13:</u> A total of $450 is divided into equal shares. If Kate receives four shares, Kevin receives three shares, and Anna receives the remaining two shares, how much money did Kevin receive?

<u>June '99, #15:</u> During a recent winter, the ratio of deer to foxes was 7 to 3 in one county of New York State. If there were 210 foxes in the county, what was the number of deer in the county?

(1) 90   (2) 147   (3) 280   (4) 490

Jan '00, #30: The volume of a rectangular pool is 1,080 cubic meters. Its length, width, and depth are in the ratio 10:4:1. Find the number of meters in each of the three dimensions of the pool.

June '01, #16: A boy got 50% of the questions on a test correct. If he had 10 questions correct out of the first 12, and ¼ of the remaining questions correct, how many questions were on the test?

(1) 16   (2) 24   (3) 26   (4) 28

June '01, #24: If a girl 1.2 meters tall casts a shadow 2 meters long, how many meters tall is a tree that casts a shadow 75 meters long at the same time?

June '01, #27: A factory packs CD cases into cartons for a music company. Each carton is designed to hold 1,152 CD cases. The Quality Control Unit in the factory expects an error of less than 5% over or under the desired packing number. What is the least number and the most number of CD cases that could be packed in a carton and still be acceptable to the Quality Control Unit?

Aug '00, #29: After an ice storm, the following headlines were reported in the *Glacier County Times*:

Monday: Ice Storm Devastates County – 8 out of every 10 homes lose electrical power
Tuesday: Restoration Begins – Power restored to ½ of affected homes
Wednesday: More Freezing Rain – Power lost by 20% of homes that had power on Tuesday

Based on these headlines, what fractional portion of homes in Glacier County had electrical power on Wednesday?

June '02, #22: Ninety percent of the ninth grade students at Richbartville High School take algebra. If 180 ninth grade students take algebra, how many ninth grade students do not take algebra?

June '02, #23: If the instructions for cooking a turkey state "Roast turkey at 325° for 20 minutes per pound," how many hours will it take to roast a 20-pound turkey at 325°?

June '02, #30: In the accompanying diagram, triangle A is similar to triangle B. Find the value of n.

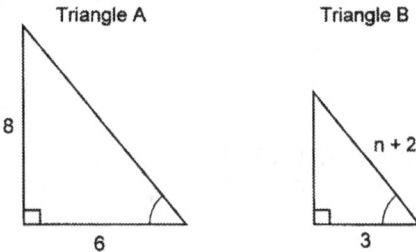

June '02, #33: Mr. Perez owns a sneaker store. He bought 350 pairs of basketball sneakers and 150 pairs of soccer sneakers from the manufacturers for $62,500. He sold all the sneakers and made a 25% profit. If he sold the soccer sneakers for $130 per pair, how much did he charge for one pair of basketball sneakers?

Aug '02, #1: On a map, 1 centimeter represents 40 kilometers. How many kilometers are represented by 8 centimeters?

(1) 5   (2) 48   (3) 280   (4) 320

Aug '02, #23: An image of a building in a photograph is 6 centimeters wide and 11 centimeters tall. If the image is similar to the actual building and the actual building is 174 meters wide, how tall is the actual building in meters?

Aug '02, #25: In bowling leagues, some players are awarded extra points called their "handicap". The "handicap" in Anthony's league is 80% of the difference between 200 and the bowler's average. Anthony's average is 145. What is Anthony's "handicap"?

Sample #33: A clothing store offers a 50% discount at the end of each week that an item remains unsold. Patrick wants to buy a shirt at the store and he says, "I've got a great idea! I'll wait two weeks, have 100% off, and get it for free!" Explain to your friend Patrick why he is incorrect and find the correct percent of discount on the original price of the shirt.

June '00, #4: Two numbers are in the ratio 2:5. If 6 is subtracted from their sum, the result is 50. What is the larger number?

(1) 55   (2) 45   (3) 40   (4) 35

Jan '03, #22: The world population was 4.2 billion people in 1982. The population in 1999 reached 6 billion. Find the percent of change from 1982 to 1999.

Jan '03, #25: Mr. Smith's class voted on their favorite ice cream flavors, and the results are shown in the accompanying diagram. If there are 20 students in Mr. Smith's class, how many students chose coffee ice cream as their favorite flavor?

Favorite Ice Cream Flavors

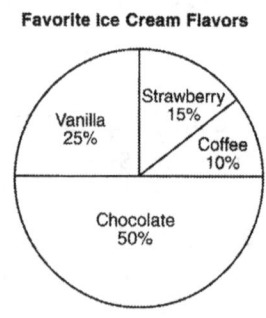

<u>Jan '03, #31:</u> At the Phoenix Surfboard Company, $306,000 in profits was made last year. This profit was shared by the four partners in the ratio 3:3:5:7. How much more money did the partner with the largest share make than one of the partners with the smallest share?

<u>June '03, #22:</u> The lengths of the sides of two similar billboards are in the ratio 5:4. If 250 square feet of material is needed to cover the larger billboard, how much material, in square feet is needed to cover the smaller billboard?

<u>June '03, #28:</u> In a town election, candidates A and B were running for mayor. There were 30,500 people eligible to vote, and $\frac{3}{4}$ of them actually voted. Candidate B received $\frac{1}{3}$ of the votes cast. How many people voted for candidate B? What percent of the votes cast, to the nearest tenth of a percent, did candidate A receive?

The Pythagorean Theorem can also be entered in SOLVER!!

Using the alphabet keys, enter the formula as:
$$A^2+B^2-C^2$$

In any Pythagorean Theorem problem you are given two of the three sides of a right triangle. If you are given the hypotenuse (the longest side, also the side opposite the right angle) enter this value for C.

Enter one or both legs (shorter sides) in A and/or B. Then solve for the missing letter.

Example:
In $\triangle ABC$, a=12, c=15, find b.

In SOLVER: | ALPHA | | MATH | | X^2 | | + | | ALPHA | | APPS | | X^2 |

| — | | ALPHA | | PRGM | | X^2 |

It should now say:
eqn: $0=A^2 + B^2 - C^2$

| ENTER |

The three variables should now be listed:
A=
B=
C=

1. Enter 12 for A.

2. Enter 15 for C.

3. Align the cursor with B.

4. Solve.　　　　B=_____

Find the missing value for each known pair: (Round decimals to the nearest hundredth.)

A. a=11,　　b=15,　　　c=_____

B. a=6,　　b=10,　　　c=_____

C. a=1,　　b=1,　　　c=_____

D. a=3,　　b=8,　　　c=_____

E. a=_____, b=11,　　　c=15

F. a=_____, b=12,　　　c=13

G. a=_____, b=7,　　　c=16

H. a=11,　　b=_____,　c=20

I. a=25,　　b=30,　　　c=_____

J. a=25,　　b=_____,　c=40

K. a=81,　　b=99,　　　c=_____

L. a=64,　　b=_____,　c=100

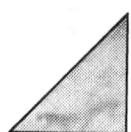

M. a=29,　　b=50,　　　c=_____

N. a=_____,　b=3,　　　c=7

Try changing "A" in (N) to a fraction. What happens?

Don't panic if a decimal will not convert to a fraction!
With any Pythagorean Theorem problem (or any other question
involving squares and/or square roots) it is possible that the
solution will be _IRRATIONAL._

An irrational number is one which CANNOT be
expressed as a fraction.

Example: Try entering π using

Then change to a fraction.

It doesn't work because π is an irrational number.

The other possibility is that the denominator is too
large.
The TI-83+/TI-84+ will not convert a decimal to a
fraction if the denominator of the fraction has more
than 3 digits.

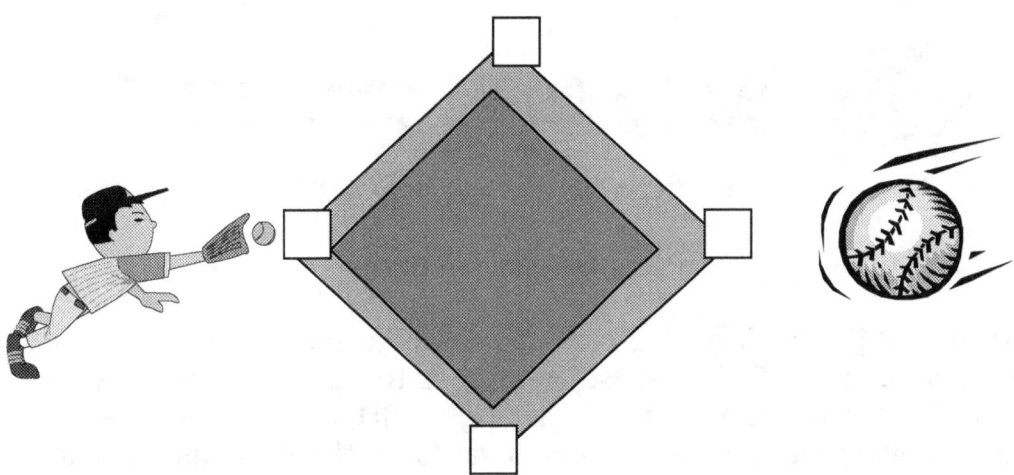

The Baseball Diamond:

The distance between each of the consecutive bases is 90 feet.

How far does a catcher have to throw the ball to get from home plate to 2nd base? How far does a 3rd baseman have to throw the ball to get from 3rd base to 1st base?

We can use the Pythagorean Theorem to help us solve this problem.

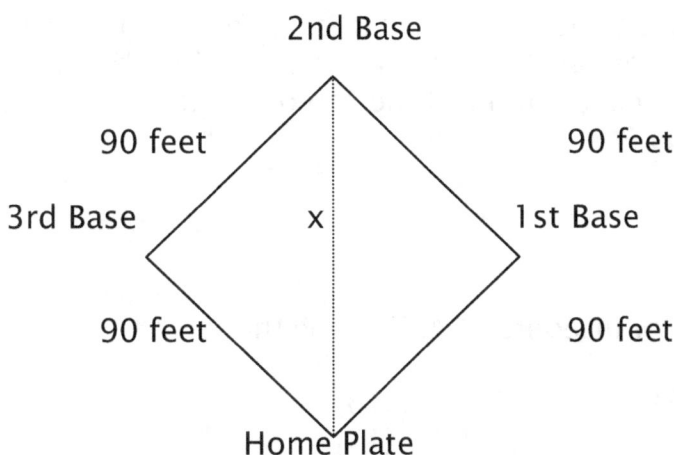

How can you use the Pythagorean Theorem to solve for x?
_____
Where is the right triangle? _____Which of the three sides is the hypotenuse? _____What are the legs? _____

Now use solver to solve the equation.   X=_____

Adapted from http://www.geom.umn.edu/~demo5337/Group3/bball.html

Regents Connection 4

These can be done with SOLVER!

Pythagorean Theorem

<u>Aug '99, #34:</u> Mr. Gonzalez owns a triangular plot of land BCD with DB=25 yards and BC=16 yards. He wishes to purchase the adjacent plot of land in the shape of right triangle ABD, as shown in the accompanying diagram, with AD=15 yards. If the purchase is made, what will be the total number of square yards in the area of his plot of land, ΔACD?

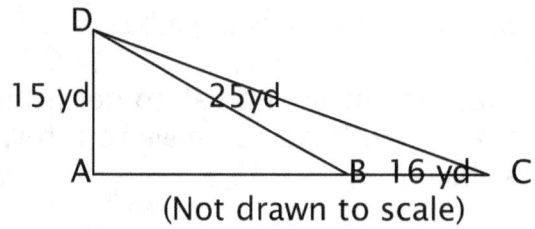

(Not drawn to scale)

<u>June '01, #15:</u> A woman has a ladder that is 13 feet long. If she sets the base of the ladder on level ground 5 feet from the side of a house, how many feet above the ground will the top of the ladder be when it rests against the house?

        (1) 8   (3) 11
        (2) 9   (4) 12

<u>June '00, #9:</u> The set of integers {3,4,5} is a Pythagorean triple. Another such set is

        (1) {6,7,8}   (3) {6,12,13}
        (2) {6,8,12}   (4) {8,15,17}

<u>Sample #11:</u> What is the distance between points A(7,3) and B(5,-1)?

    (1) $\sqrt{10}$ (2) $\sqrt{12}$ (3) $\sqrt{14}$ (4) $\sqrt{20}$

Sample '98, #34: A 10-foot ladder is placed against the side of a building as shown in figure 1 below. The bottom of the ladder is 8 feet from the base of the building. In order to increase the reach of the ladder against the building, it is moved 4 feet closer to the base of the building as shown in figure 2.

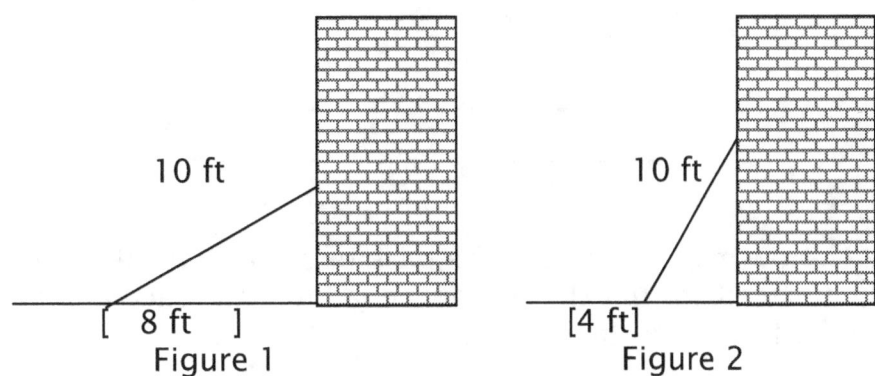

Figure 1   Figure 2

To the nearest foot, how much further up the building does the ladder now reach? Show how you arrived at your answer.

Jan '00, #23: A wall is supported by a brace 10 feet long, as shown in the diagram below. If one end of the brace is placed 6 feet from the base of the wall, how many feet up the wall does the brace reach?

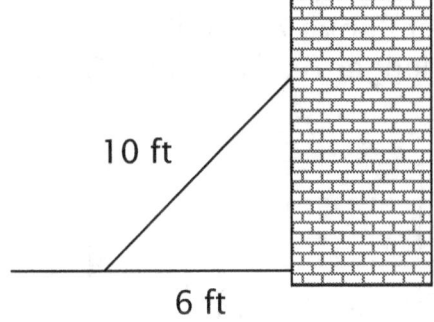

Aug '00, #30: Katrina hikes 5 miles north, 7 miles east, and then 3 miles north again. To the nearest tenth of a mile, how far, in a straight line, is Katrina from her starting point?

<u>June '99, #33:</u> The cross section of an attic is in the shape of an isosceles trapezoid, as shown in the accompanying figure. If the height of the attic is 9 feet, BC=12 feet, and AD=28 feet, find the length of AB to the nearest foot.

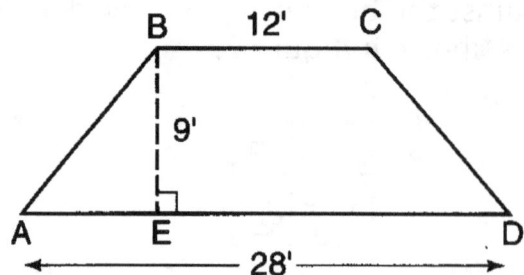

<u>Aug '01, #22:</u> How many feet from the base of a house must a 39-foot ladder be placed so that the top of the ladder will reach a point on the house 36 feet from the ground?

<u>Jan '02, #2:</u> If the lengths of the legs of a right triangle are 5 and 7, what is the length of the hypotenuse?

(1) $\sqrt{2}$ (2) $2\sqrt{3}$ (3) $2\sqrt{6}$ (4) $\sqrt{74}$

<u>Jan '02, #26:</u> Jerry and Jean Jogger start at the same time from point A shown on the accompanying set of axes. Jerry jogs at a rate of 5 miles per hour traveling from point A to point R to point S and then to point C. Jean jogs directly from point A to point C on $\overline{AC}$ at the rate of 3 miles per hour. Which jogger reaches point C first? Explain or show your reasoning.

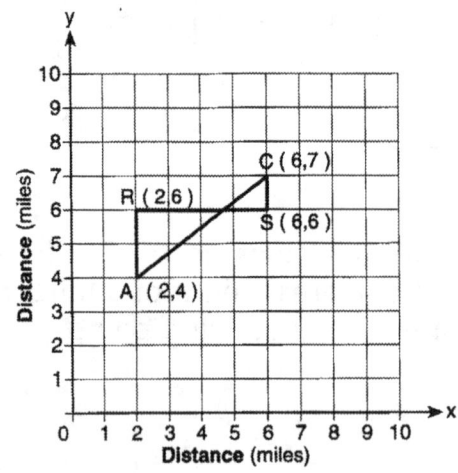

<u>Jan '03, #30:</u> A rectangular garden is going to be planted in a person's rectangular back yard, as shown in the accompanying diagram. Some dimensions of the back yard and the width of the garden are given. Find the area of the garden to the nearest square foot.

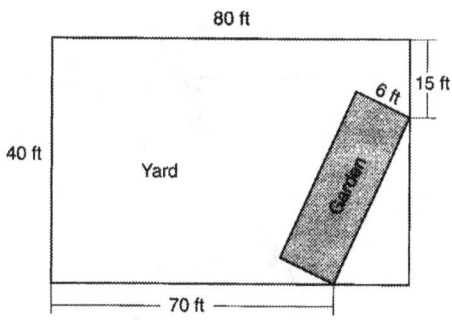

<u>June '03, #30:</u> To get from his high school to his home, Jamal travels 5.0 miles east and then 4.0 miles north. When Sheila goes to her home from the same high school, she travels 8.0 miles east and 2.0 miles south. What is the measure of the shortest distance, to the nearest tenth of a mile, between Jamal's home and Sheila's home" [The use of the accompanying grid is optional.]

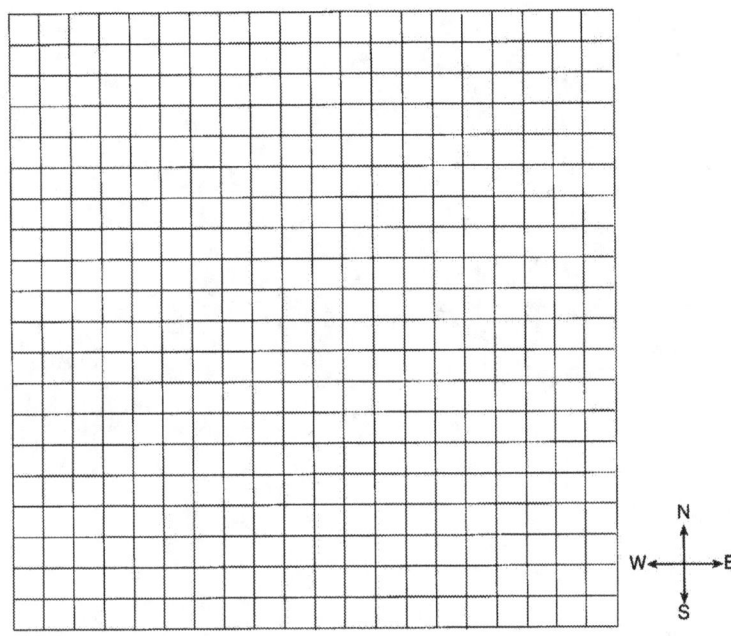

<u>June '03, #34:</u> A straw is placed into a rectangular box that is 3 inches by 4 inches by 8 inches, as shown in the accompanying diagram. If the straw fits exactly into the box diagonally from the bottom left front corner to the top right back corner, how long is the straw, to the nearest tenth of an inch?

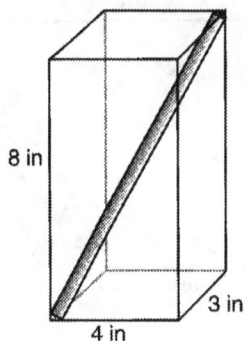

LCM ~ GCF

The LCM (Least Common Multiple) or the GCF (Greatest Common Factor) of two numbers can both be done on the graphing calculator.

Suppose you want to know the LCM of 8 and 28.

1. | MATH |

2. ⟶ NUM

3.   Choose   8:lcm(

4.   Enter the two numbers with a comma

               between them: lcm(8,28

5. | ENTER | The LCM=_____

The calculator will not accept more than two numbers when finding the LCM or GCF. To find the LCM of three numbers, find the LCM of the first two, then find the LCM of your answer and the third number.

Find the GCF of 24, 36, 96

1. | MATH |

2. ⟶ NUM

3. Choose 9:gcd(

4. Enter the first two numbers with a comma between them:

          gcd(24,36

5. | ENTER | the GCF of 24 and 36 is 12.

6. Repeat from 1-3. Enter the answer from above (12) and the third number from the original problem: gcd(12,96

7. ENTER . The GCF of 24,36,96 is _____ .

***Notice that the calculator calls the GCF gcd** instead. They are simply using the word _____ in place of the word _____. They mean exactly the same thing!

Find the LCM:
1. 5,35   _____
2. 8,18   _____
3. 3,9,27   _____
4. 7,5,21   _____
5. 8,19,3   _____
6. 16,18,20   _____
7. 22,33,44   _____
8. 15,35,8   _____
9. 21,33,14   _____
10. 12,13,14,15   _____

Find the GCF:
1. 15,35   _____
2. 18,56   _____
3. 48,32   _____
4. 88,96   _____
5. 108,96,144   _____
6. 112,104,92   _____
7. 100,255,315   _____
8. 121,48,60   _____
9. 130,65,195,260   _____
10. 256,316,204,366   _____

Evaluating Expressions & Absolute Value

The graphing calculator can evaluate algebraic expressions if the value of the variable(s) is identified.

The first step is very important!! Recall that the calculator retains the previously used value for any variable. Therefore, the first step is to assign the new value to the variable you plan to use.

Example: What is the value of x^3-4x^2 for $x=-5$?

1. To change the value of x to –5:

   | - | | 5 | | STO→ | | X,T,θ,n |

2. There are two ways to evaluate the expression from this point.

   a. Without pressing enter after storing, follow the store key with a colon (:) then key in the expression. Then press ENTER.

   OR

   b. ENTER after storing. Then key in the expression and ENTER again. (This is usually the easier choice!)

3. x^3-4x^2 = _____

If the expression includes an absolute value this can be found by pressing MATH, moving over to NUM and choosing #1, abs(.

Example: What is the value of $\left| x^3 - 18 \right|$ for x=-4?

| (-) | 4 | STO→ | X,T,θ,n | ALPHA | • |

MATH ⟶ NUM   Choose 1:abs(| X,T,θ,n |

| ^ | 3 | - | 1 | 8 |) | ENTER |

$\left| x^3 - 18 \right|$ = _____ when x=4

Practice:

1. $x^2 - 5x + 7$, for x=-6   _____

2. $-x^4 + 8$, for x=3   _____

3. $10x^3 + 7$, for x=-5   _____

4. $x^4 - 2x^3 + 3x - 7$, for x=3   _____

5. $\left| -x^3 \right|$, for x=6   _____

6. $\left| 2x^2 - 8x + 15 \right|$, for x=-7   _____

7. $2x + \left| 3x^2 - 18 \right| - 10$, for x=-2   _____

8. $\left| 14x - 3 \right| + 9$, for x=-8   _____

9. $114 - \left| x^5 - 20 \right| + 7$, for x=-3   _____

10. $\left| x^4 - 3 \right| - \left| x^3 + 7 \right| + 12$, for x=3   _____

Regents Connection 5

Evaluating Expressions

<u>Jan '00, #15:</u> If t=-3, then $3t^2+5t+6$ equals

    (1) –36   (2) –6   (3) 6   (4) 18

<u>Aug '99, #8:</u> The formula $C=\dfrac{5}{9}(F-32)$ can be used to find the Celsius temperature (C) for a given Fahrenheit temperature (F). What Celsius temperature is equal to a Fahrenheit temperature of 77°?

    (1) 8°   (2) 25°   (3) 45°   (4) 171°

<u>June '00, #21:</u> the formula for changing Celsius (C) temperature to Fahrenheit (F) temperature is $F=\dfrac{9}{5}C+32$. Calculate, to the nearest degree, the Fahrenheit temperature when the Celsius temperature is – 8.

Inequalities and Logic

Inequalities:

The graphing calculator will test an inequality for you to see if it is a true statement.

Example:
Given the question "Which members of the domain {-2,-5,10} are solutions of the open sentence $2(x + 5) > 4x + 9$?"

Step 1:
The first value can be inserted for X by entering | (-) | | 2 | then

| STO ⇒ | | X,T,θ,n | Press | ENTER |

Step 2:
Enter the sentence on the home screen of the calculator. (The

inequality symbol is found under the "TEST" menu. Use | 2nd |,

| MATH |.) Press | ENTER | when finished.

The result will be either the number 1, or the number 0. If it is **1**, then the inequality was **true**. If it is **0**, the inequality is **false**.

Step 3:
To try the next value, repeat step 2 using the next value instead 0f-2.

Step 4:
Press | 2nd | | ENTER | | 2nd | | ENTER |

This should bring back your inequality. Press | ENTER | The result will tell whether the inequality is true or false with the new value entered.

Repeat with the remaining value

Practice:

Test the following expressions. (True or False)

1. $7^2 - 5^2 = 24$ _____

2. $4 \cdot 7 + 3 = 40$ _____

3. 5% of $180 = 90$ (remember that "of" translates as multiplication!) _____

4. $-9 > -12$ _____

5. $3x^2 < 14$, for x=-4 _____

6. $x^2 - 3x + 2 = (x-2)(x+1)$ _____

7. $x^2 + 4x + 4 = (x+2)^2$ _____

Logic:

The graphing calculator can be used to find whether a conjunction or disjunction is true or false.

Example:

The conjunction $(5<6) \land (9=5 + 4)$ can be entered in the home screen just as it is written. The < symbol can be found in the TEST menu and the $\land$ symbol is in the LOGIC menu as the word "and".

Recall, the TEST menu is found by using [2nd] [MATH] .

The LOGIC menu is in the same place; just use the right arrow key to move the cursor over to LOGIC. Choose "1" for conjunctions, "2" for disjunctions.

When you have finished keying in the conjunction, use the [ENTER] key.

As with the inequalities, the result will be either a "1" or a "0".

Practice:

Test the following conjunctions and disjunctions. Write whether they are true or false.

1. $(7>5) \lor (4=6)$ _____

2. $(12 - 6=6) \lor (7 \neq 2)$ _____

3. $(5 > -1) \land (-8 < -3)$ _____

4. $(-4^2 = -16) \land (4^2 < 2^4)$ _____

5. $(26 = 3^3 - 1) \lor (6 > 5 \cdot 3)$ _____

6. $(11 = 4^2 - 5) \lor (-5 > 3)$ _____

7. $(10 - 5 \cdot 2 < 0) \land (4^3 \neq 64)$ _____

8. $(14 = 5 \cdot 3 - 1) \land (4 + 3^2 < 13)$ _____

Inequalities

<u>June '99, #17:</u> If $t^2 < t < \sqrt{t}$, then t could be

(1) -¼   (2) 0   (3) ¼   (4) 4

<u>June '99, #28:</u> A swimmer plans to swim at least 100 laps during a 6-day period. During this period, the swimmer will increase the number of laps completed each day by one lap. What is the least number of laps the swimmer must complete on the first day?

<u>June '01, #18:</u> In the set of positive integers, what is the solution set of the inequality 2x-3<5?

(1) {0,1,2,3}   (3) {0,1,2,3,4}
(2) {1,2,3}   (4) {1,2,3,4}

<u>Jan '01, #1:</u> There are 461 students and 20 teachers taking buses on a trip to a museum. Each bus can seat a maximum of 52. What is the least number of buses needed for the trip?

(1) 8   (2) 9   (3) 10   (4) 11

<u>Jan '03, #4:</u> In which list are the numbers in order from least to greatest?

(1) 3.2, π, $3\frac{1}{3}$, $\sqrt{3}$ (3) $\sqrt{3}$, π, 3.2, $3\frac{1}{3}$

(2) $\sqrt{3}$, 3.2, π, $3\frac{1}{3}$ (4) 3.2, $3\frac{1}{3}$, $\sqrt{3}$, π

<u>Jan '03, #12:</u> Which graph represents the solution set for $2x - 4 \leq 8$ and $x + 5 \geq 7$?

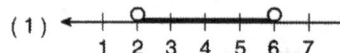

(1) ← | ○━━━━━○ | →
     1 2 3 4 5 6 7

(2) ← | ●━━━━━● | →
     1 2 3 4 5 6 7

(3) ←━━━○ | | | ○━━━→
     1 2 3 4 5 6 7

(4) ←━━━● | | | ●━━━→
     1 2 3 4 5 6 7

<u>June '03, #11:</u> Which number is in the solution set of the inequality $5x + 3 > 38$?

(1) 5   (2) 6   (3) 7   (4) 8

Logic

<u>June '99, #2:</u> The statement "If x is divisible by 8, then it is divisible by 6" is false if x equals

(1) 6   (2) 14   (3) 32   (4) 48

<u>Jan '00, #3:</u> Mary says, "The number I am thinking of is divisible by 2 or it is divisible by 3." Mary's statement is false if the number she is thinking of is

(1) 6   (2) 8   (3) 11   (4) 15

<u>Jan '00, #25:</u> Al says, "If ABCD is a parallelogram, then ABCD is a rectangle." Sketch a quadrilateral ABCD that shows that Al's statement is not always true. Your sketch **must show the length of each side and the measure of each angle** for the quadrilateral you draw.

<u>June '01, #12:</u> Which statement is logically equivalent to "If I eat, then I live"?

(1)   If I live, then I eat.
(2)   If I eat, then I do not live.
(3)   I live if and only if I eat.
(4)   If I do not live, then I do not eat.

<u>Aug '99, #12:</u> What is true about the statement "If two angles are right angles, the angles have equal measure; and its converse "If two angles have equal measure then the two angles are right angles"?

   (1)   The statement is true but its converse is false.
   (2)   The statement is false but its converse is true.
   (3)   Both the statement and its converse are false.
   (4)   Both the statement and its converse are true.

<u>Jan '01, #12:</u> Given the statement: "If two sides of a triangle are congruent, then the angles opposite these sides are congruent."

Given the converse of the statement: "If two angles of a triangle are congruent, then the sides opposite these angles are congruent."

What is true about this statement and its converse?

   (1)   Both the statement and its converse are true.
   (2)   Neither the statement nor its converse is true.
   (3)   The statement is true but its converse is false.
   (4)   The statement is false but its converse is true.

<u>Jan '01, #29:</u> Mark says, "The number I see is odd." Jan says, "That same number is prime." The teacher says, "Mark is correct or Jan is correct." Some integers would make the teacher's statement true while other integers would make it false. Give **and explain** one example of when the teacher's statement is **true**. Give **and explain** one example of then the teacher's statement is **false**.

<u>Aug '00, #14:</u> What is the converse of the statement "If it is sunny, I will go swimming"?

   (1)   If it is not sunny, I will not go swimming.
   (2)   If I do not go swimming, then it is not sunny.
   (3)   If I go swimming, it is sunny.
   (4)   I will go swimming if and only if it is sunny.

Aug '00, #26: John, Dan, Karen, and Beth went to a costume ball. They chose to go as Anthony and Cleopatra, and Romeo and Juliet. John got the costumes for Romeo and Cleopatra, but not his own costume. Dan saw the costumes for Juliet and himself. Karen went as Anthony. Beth drove two of her friends, who were dressed as Anthony and Cleopatra, to the ball. What costume did John wear?

Aug '01, #4: Which statement is logically equivalent to "If I did not eat, then I am hungry"?

    (1)   If I am not hungry, then I did not eat.
    (2)   If I did not eat, then I am not hungry.
    (3)   If I am not hungry, then I did eat.
    (4)   If I am hungry, then I did eat.

Aug '01, #16: Which statement is the converse of "If it is a 300 ZX, then it is a car"?

    (1)   If it is not a 300 ZX, then it is not a car.
    (2)   If it is not a car, then it is not a 300 ZX.
    (3)   If it is a car, then it is a 300 ZX.
    (4)   If it is a car, then it is not a 300 ZX.

June '02, #21: Given the statement "John is not handsome" and the false statement "John is handsome or smart." Determine the truth value for the statement "John is smart." (A true or false response with no explanation receives no credit!!!)

Aug '02, #5: Given the statement: "If two lines are cut by a transversal so that the corresponding angles are congruent, then the lines are parallel."

What is true about the statement and its converse?

    (1) The statement and its converse are both true.
    (2) The statement and its converse are both false.
    (3) The statement is true, but its converse is false.
    (4) The statement is false, but its converse is true.

<u>Jan '02, #14:</u> Frank, George, and Hernando are a plumber, a cabinetmaker, and an electrician, though not necessarily in that order. Each can do all work appropriate to his own field, but no work in other fields. Frank was not able to install a new electric line in his home. Hernando was not able to make cabinets. George is also a building contractor who hired one of the other people to do his electrical work. Which statement must be true?

- (1)   Hernando is an electrician.
- (2)   George is a cabinetmaker.
- (3)   Frank is a plumber.
- (4)   Frank is an electrician.

<u>Jan '02, #20:</u> Which statement is logically equivalent to "If the team has a good pitcher, then the team has a good season"?

- (1) If the team does not have a good season, then the team does not have a good pitcher.
- (2) If the team does not have a good pitcher, then the team does not have a good season.
- (3) If the team has a good season, then the team has a good pitcher.
- (4) The team has a good pitcher and the team does not have a good season.

<u>Sample #12:</u> "If Mary and Tom are classmates, then they go to the same school." Which statement below is logically equivalent?

- (1) If Mary and Tom do not go to the same school, then they are not classmates.
- (2) If Mary and Tom are not classmates, then they do not go to the same school.
- (3) If Mary and Tom go to the same school, then they are classmates.
- (4) If Mary and Tom go to the same school, then they are not classmates.

<u>June '00, #6:</u> What is the inverse of the statement "If it is sunny, I will play baseball?"

- (1) If I play baseball, then it is sunny.
- (2) If it is not sunny, I will not play baseball.
- (3) If I do not play baseball, then it is not sunny.
- (4) I will play baseball if and only if it is sunny.

Jan '03, #3: What is the inverse of the statement "If Mike did his homework, then he will pass this test"?

(1) If Mike passes this test, then he did his homework.
(2) If Mike does not pass this test, then he did not do his homework.
(3) If Mike does not pass this test, then he only did half his homework.
(4) If Mike did not do his homework, then he will not pass this test.

Jan '03, #8: Given the true statement: "If a person is eligible to vote, then that person is a citizen."

Which statement must also be true?

(1) Kayla is not a citizen; therefore, she is not eligible to vote.
(2) Juan is a citizen; therefore, he is eligible to vote.
(3) Marie is not eligible to vote; therefore, she is not a citizen.
(4) Morgan has never voted; therefore, he is not a citizen.

June '03, #8: Which statement is logically equivalent to "If it is Saturday, then I am not in school?

(1) If I am not in school, then it is Saturday.
(2) If it is not Saturday, then I am in school.
(3) If I am in school, then it is not Saturday.
(4) If it is Saturday, then I am in school.

June '03, #17: What is the inverse of the statement "If Julie works hard, then she succeeds"?

(1) If Julie succeeds, then she works hard.
(2) If Julie does not succeed, then she does not work hard.
(3) If Julie works hard, then she does not succeed.
(4) If Julie does not work hard, then she does not succeed.

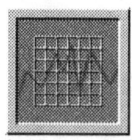

Statistics

The graphing calculator is extremely useful in organizing and comparing data.

While most people do not find the statistics necessary for the Math A exam difficult, the graphing calculator can save valuable time that might be better used on other questions.

When beginning any statistics based question you will be provided with a set of data.

Your first task will be to enter this data into the calculator as a list.

There is more than one way to enter a list, but we will use only the list editor. This is found by using | STAT | , choose

1:Edit... , and enter the data in an available list.

Example:
10 students have taken a Math A quiz. Their scores are:
85,92,73,77,82,85,95,100,60,75

1. Go to | STAT | .

2.   Choose 1:Edit... from the Edit menu.

3.   If L_1 is empty enter the data in the list by keying in the number and pressing | ENTER | .

   Be sure to press ENTER or the down arrow after **each** number in the list **(even the last one!)**

4. If L_1 is not empty clear it by pressing the up arrow until the name of the list is highlighted. Press | CLEAR | then

   | ENTER | . Then enter the data in the list.

5. **Do not use the delete key to empty the list**. This will delete the whole list including its name. Resetting the calculator will restore lists L_1 through L_6.

When the list is complete you can have the calculator arrange the data in ascending (SortA) or descending order (SortD).

1. Press │ STAT │

2. Choose 2:SortA(

3. │ 2nd │ │ 1 │ will place L_1 after SortA(on the home

   screen. │ ENTER │

4. This will not take you back to the list. To check the list you must return to the Edit menu in STAT.

5. When the list has been sorted it can be quickly scanned using the up and down arrow keys to determine if there is a mode. What is the mode for the quiz scores?_____

6. Next we will have the calculator calculate other

   information for us. Press│ STAT │, use the right arrow over to

   CALC, then choose 1:1-Var Stats. Enter the list name (L_1).

7. You should now have a list from which you can fill in the following table.

Mean ($\bar{x}$)	
n (how many)	
minX	
Q1	
Med (median)	
Q3	
maxX	

8. Using the sorted list make a stem-and-leaf plot for the data.

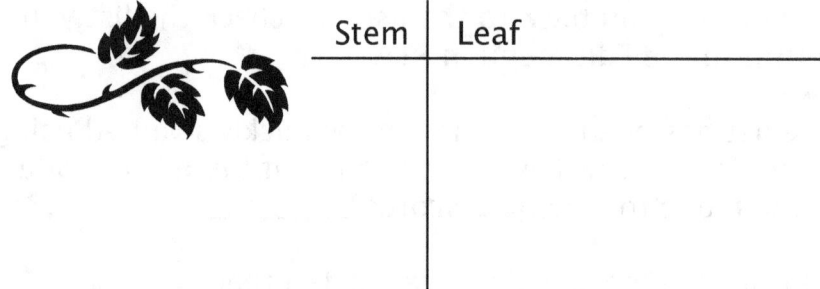

Stem	Leaf

9. Using the table above, make a box-and-whisker plot for the data.

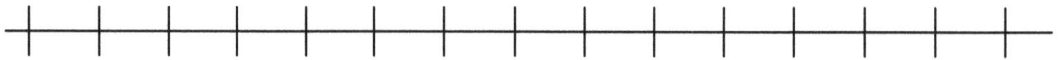

Histograms

Histograms are a common way to display data that can be grouped into _____ .

Consider the test scores for Test 1 in Graphing Calculator:

48, 92, 93, 86, 62, 78, 66, 66, 90, 90, 90, 50, 67, 95, 90, 98, 90, 100, 70, 77, 100, 54, 27, 89, 95, 100, 98, 88, 83, 73, 90, 77, 78, 64, 82, 61, 75, 95, 82, 90, 74, 93, 19, 90, 94

Begin by entering the data in L_1.
Sort list 1.
Complete the table below:

Interval	Frequency	Cumulative Frequency
0-10		
11-20		
21-30		
31-40		
41-50		
51-60		
61-70		
71-80		
81-90		
91-100		

What is the difference between frequency and cumulative frequency?
_____
_____
_____

What distinguishes a histogram from a bar graph?
_____
_____

Create a properly labeled frequency histogram of the test scores on the grid below.

Create a cumulative frequency histogram of the test scores below.
Label appropriately!

What can be said about the test scores based upon the graphs?

To check your frequency histogram you can plot the graph on the graphing calculator.

1. Press 2nd Y= (Stat Plot).

2. Press ENTER

3. Using the arrow keys highlight ON and press ENTER .

4. Highlight the picture of the type of graph we want and press

   ENTER .

5. Be sure it says L₁ after Xlist:

6. Press GRAPH

```
          Plot2  Plot3
On Off
Type:
Xlist:L1
Freq:1
```

7. The default screen will probably not give the best view (if any)

   of the data. To see a better graph, press ZOOM , choose

   9:ZoomStat. Now the calculator will adjust the screen based on

   the type of plot you have chosen and the range of data.

8. Sketch the screen in the box below.

Does the screen match the graph you made on page 68?

Press [WINDOW] . Your screen should look like the one below.

```
WINDOW
 Xmin=■9
 Xmax=113.5
 Xscl=13.5
 Ymin=-5.71311
 Ymax=22.23
 Yscl=1
 Xres=1
```

The calculator has decided where to begin (Xmin) and the size of the intervals (Xscl). Who ever heard of a histogram with intervals of 13.5?!!

Easily changed! **With the cursor on Xmin enter 1** (starting at 0 will throw off our intervals unfortunately.) Then **with the cursor on Xscl enter 10** so that it now looks like the screen below.

```
WINDOW
 Xmin=1
 Xmax=113.5
 Xscl=10
 Ymin=-5.71311
 Ymax=22.23
 Yscl=1
 Xres=1
```

Now press [GRAPH] . Does your graph look like the one below?

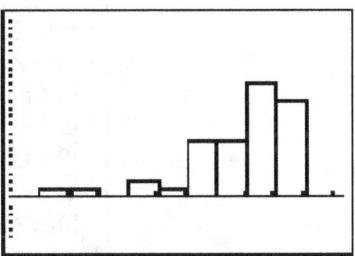

To check the intervals and frequency press TRACE and use the left and right arrow keys. The values on the screen below tell us that the cursor is in the interval 61-70 and the frequency for this interval is 7.

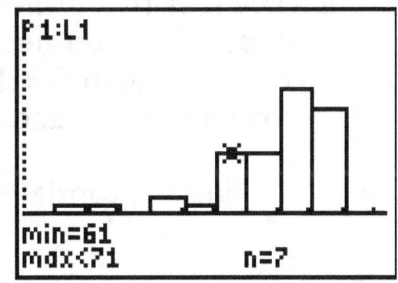

A measure of dispersion shows how the data is spread out.

Box-and-Whisker plots are good pictures of dispersion.

Easy Box-and-Whisker Plots:

1. Enter data in L_1.
   Quiz Scores: 78, 90, 85, 72, 100,
   84, 60, 85, 95, 55, 75, 88, 97,
   74, 77, 80, 92, 69, 75, 83

2. Press STAT

3. Move over to CALC

4. Choose 1:1-Var Stats

5. ENTER (Unless you have more than one list. If there is more
   than one list you need to name the list you need the statistics
   for before pressing ENTER.)

6. If you need the mean it is the first entry on the list.

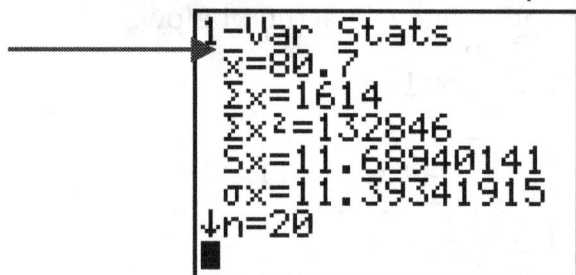

```
1-Var Stats
x̄=80.7
Σx=1614
Σx²=132846
Sx=11.68940141
σx=11.39341915
↓n=20
■
```

7. Otherwise, scroll down until you see:

Minimum
First (or Lower) Quartile
Median
Third (or Upper) Quartile
Maximum

```
1-Var Stats
↑n=20
minX=55
Q₁=74.5
Med=81.5
Q₃=89
maxX=100
■
```

8. **The last five numbers are all you need and they are in the
   order that you need them.**
   Plot the minimum, first quartile, median, third quartile, and
   maximum just above the number line.

9. To find appropriate intervals, find the range of the data and

divide by the number of intervals. Round up to a "nice" number.
(i.e. a multiple of 2, 5, or 10 is usually easy to work with.)

10.   Make vertical lines at the middle three points and make them into a box.

11.   Make "whiskers" extending out to the min and max.

12.   To check your work you can see what the plot should look like by creating a STAT PLOT.

    a. Press 2nd , Y=.

    b. Turn on Plot 1 (Pressing ENTER twice should do it unless it is already turned on.)

    c. Use the arrow keys (Down then left or right.) to highlight the box-and-whisker plot. Be sure to choose the one with the vertical line inside the box at the median. Press ENTER.

    d. Be sure the Xlist matches the list name where your data was entered.

    e. Press ZOOM and choose 9:ZoomStat. This will automatically fit your data to the type of plot you have chosen. ***If an INVALID DIM Error occurs the list you have entered as the Xlist probably has no data. Check your list name.

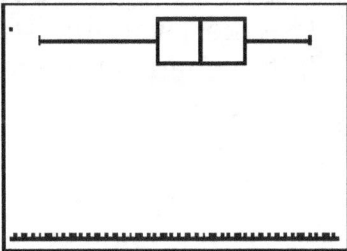

Deer Take in Lewis County, NY
2000

Town	Bucks	Total
Croghan	411	604
Denmark	163	412
Diana	212	289
Greig	132	182
Harrisburg	101	246
Lewis	95	110
Leyden	77	107
Lowville	101	209
Lyonsdale	129	169
Martinsburg	176	238
Montague	183	227
New Bremen	107	211
Osceola	195	236
Pinckney	114	237
Turin	118	163
Watson	175	268

Source: www.dec.state.ny.us

1. Enter bucks taken in L_1.
2. Enter Total (for each town) in L_2.
3. Use the Sort A(command to arrange the lists.
4. Find the mode for each set of data if one exists and enter in the chart on the next page.
5. Use the 1-variable statistics command on first L_1, then L_2 and fill in the chart on the next page.

	Bucks	All Deer
Mode		
Median		
Mean		
Minimum x		
1st Quartile		
3rd Quartile		
Maximum x		

6. Create a box-and-whisker plot for bucks taken using the information in the table above.

7. Create a box-and-whisker plot for all deer taken in Lewis County using the table above.

Statistics Practice

Using the data sheet provided and the graphing calculator:

1. Give the mean, median, mode, and range for syrup production in New York State.

    Mean= _____
    Median=_____
    Mode= _____
    Range= _____

2. Create a stem-and-leaf plot for this data.

    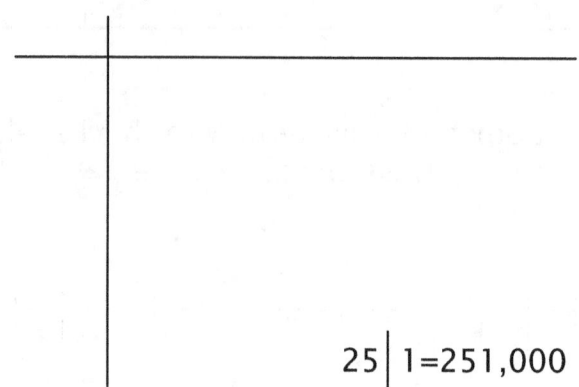

    25│1=251,000

3. Give the mean, median, mode, and range for US Maple Syrup production.

    Mean= _____
    Median=_____
    Mode= _____
    Range= _____

4. Create a histogram for this data using intervals of 1,000,000 liters. Label appropriately!

New York State Maple Syrup Production

Year	Gallons
1994	251,000
1995	208,000
1996	343,000
1997	269,000
1998	239,000
1999	195,000
2000	210,000
2001	193,000

Source: New England Agricultural Statistics Service and NYS Ag and Markets

U.S. Maple Syrup Production

Year	Liters
1984	5,106,000
1985	4,979,000
1986	3,874,000
1987	3,429,000
1988	4,557,000
1989	4,687,000
1990	4,347,000
1991	4,985,000
1992	6,211,000
1993	3,812,000
1994	5,011,000
1995	4,143,000
1996	5,890,000
1997	4,884,000
1998	4,378,000

Source: USDA

State Populations

Using the data provided:

1. Enter the state populations into L_1. You should notice that when the entry in the list exceeds 6 digits, the calculator converts the value to scientific notation in the list, but also note that the original value is still the one that appears at the bottom of the screen when the cursor is on a particular entry.

2. Sort L_1.

3. Make a histogram on graph paper or a grid using intervals of 1,000,000. Be sure to label all parts! The last interval may contain all those greater than 10,000,000.

4. What does the histogram tell us about state populations? Explain in detail using complete sentences.

US Population By State

State:	Population:	State:	Population:
Alabama	4,447,100	Montana	902,195
Alaska	626,932	Nebraska	1,711,263
Arizona	5,130,632	Nevada	1,998,257
Arkansas	2,673,400	New Hampshire	1,235,786
California	33,871,648	New Jersey	8,414,350
Colorado	4,301,261	New Mexico	1,819,046
Connecticut	3,405,565	New York	18,976,457
Delaware	783,600	North Carolina	8,049,313
District of Columbia	572,059	North Dakota	642,200
Florida	15,982,059	Ohio	11,353,140
Georgia	8,186,453	Oklahoma	3,450,654
Hawaii	1,211,537	Oregon	3,421,399
Idaho	1,293,953	Pennsylvania	12,281,054
Illinois	12,419,293	Rhode Island	1,048,319
Indiana	6,080,485	South Carolina	4,012,012
Iowa	2,926,324	South Dakota	754,844
Kansas	2,688,418	Tennessee	5,689,283
Kentucky	4,041,769	Texas	20,851,820
Louisiana	4,468,976	Utah	2,233,169
Maine	1,274,923	Vermont	608,827
Maryland	5,296,486	Virginia	7,078,515
Massachusetts	6,349,097	Washington	5,894,121
Michigan	9,938,444	West Virginia	1,808,344
Minnesota	4,919,479	Wisconsin	5,363,675
Mississippi	2,844,658	Wyoming	493,782
Missouri	5,595,211		

Source: 2000 Census

Jan '01, #32: On a science quiz, 20 students received the following scores: 100, 95, 95, 90, 85, 85, 85, 80, 80, 80, 80, 75, 75, 75, 70, 70, 65, 65, 60, 55.

Construct a statistical graph, such as a histogram or a stem-and-leaf plot, to display this data. [Be sure to title the graph and label all axes or parts used.]

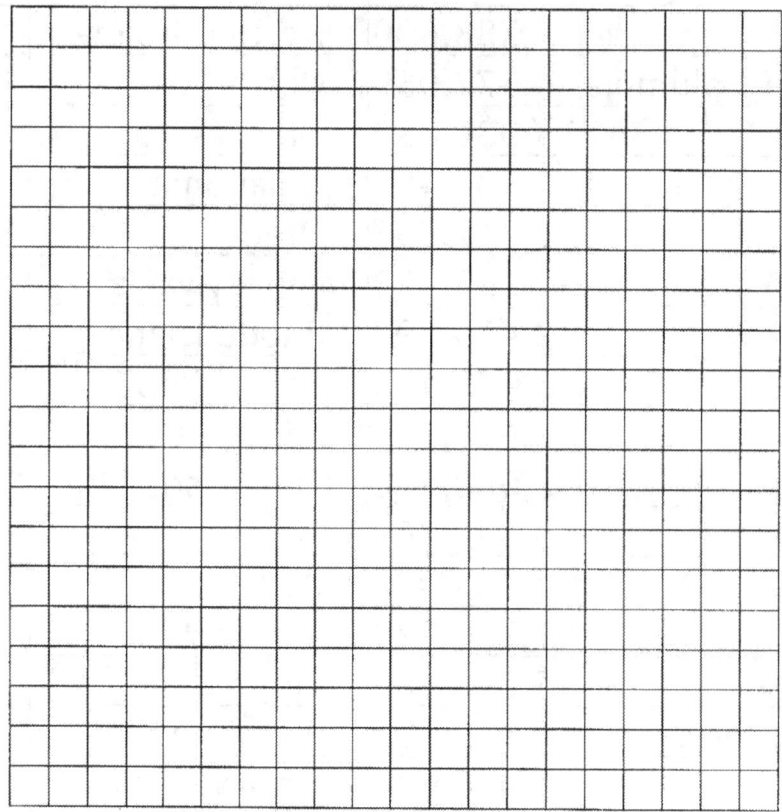

<u>Aug '01, #34:</u> The following data consists of the weights, in pounds, of 30 adults:

195, 206, 100, 98, 150, 210, 195, 106, 195, 168, 180, 212, 104, 195, 100, 216, 195, 209, 112, 99, 206, 116, 195, 100, 142, 100, 135, 98, 160, 155

Using the data, complete the accompanying cumulative frequency table and construct a cumulative frequency histogram on the grid below.

Interval	Frequency	Cumulative Frequency
51-100		
101-150		
151-200		
201-250		

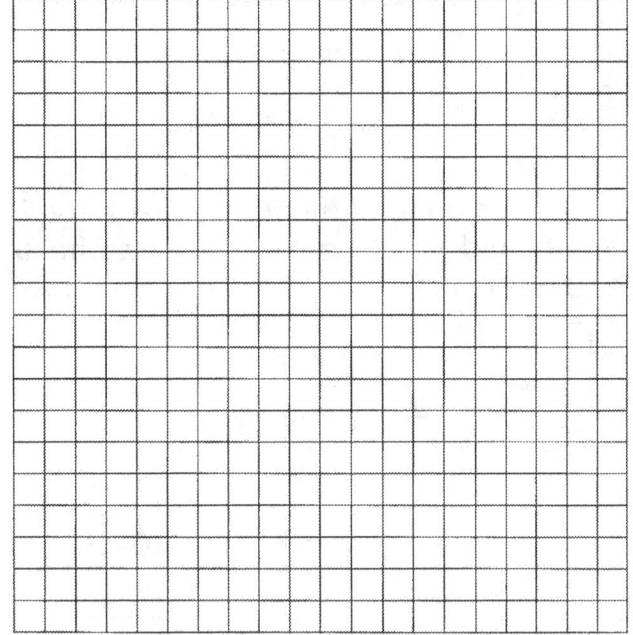

<u>Jan '02, #30:</u> The students in Woodland High School's meteorology class measured the noon temperature every school day for a week. Their readings for the first 4 days were Monday, 56°; Tuesday, 72°; Wednesday, 67°; and Thursday, 61°. If the mean (average) temperature for the 5 days was exactly 63°, what was the temperature on Friday?

Jan. '00, #5: What was the median high temperature in Middletown during the 7-day period shown in the table below?

Daily High Temperature in Middletown	
Day	Temp (°F)
Sunday	68
Monday	73
Tuesday	73
Wednesday	75
Thursday	69
Friday	67
Saturday	63

(1) 69   (3) 73
(2) 70   (4) 75

June '00, #33: The scores on a mathematics test were 70, 55, 61, 80, 85, 72, 65, 40, 74, 68, and 84. Complete the accompanying table, and use the table to construct a frequency histogram for these scores.

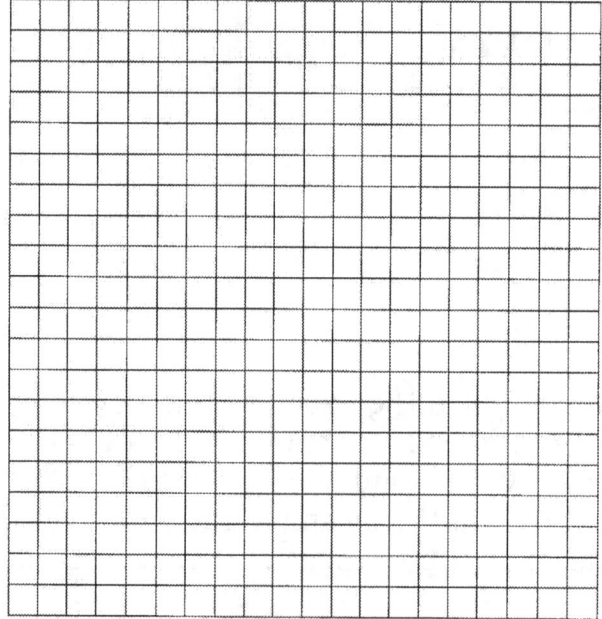

Score	Tally	Frequency
40-49		
50-59		
60-69		
70-79		
80-89		

June '99, #29: The mean (average) weight of three dogs is 38 pounds. One of the dogs, Sparky, weighs 46 pounds. The other two dogs, Eddie and Sandy, have the same weight. Find Eddie's weight.

Jan '00, #26: Judy needs a mean (average) score of 86 on four tests to earn a midterm grade of B. If the mean of her scores for the first three tests was 83, what is the lowest score on a 100-point scale that she can receive on the fourth test to have a midterm grade of B?

Jan '01, #18: From January 3 to January 7, Buffalo recorded the following daily high temperatures: 5°, 7°, 6°, 5°, and 7°. Which statement about the temperatures is true?

    (1) mean=median   (3) median=mode
    (2) mean=mode   (4) mean<median

Aug '00, #8: On an English examination, two students received scores of 90, five students received 85, seven students received 75, and one student received 55. The average score on this examination was

    (1) 75   (2) 76   (3) 77   (4) 79

Aug '01, #10: The exact average of a set of six test scores is 92. Five of these scores are 90, 98, 96, 94, and 85. What is the other test score?

    (1) 92   (2) 91   (3) 89   (4) 86

Aug '99, #10: On June 17, the temperature in New York City ranged from 90° to 99°, while the temperature in Niagara Falls ranged from 60° to 69°. The difference in the temperatures in these two cities must be between

    (1) 20° and 30°   (3) 25° and 35°
    (2) 20° and 40°   (4) 30° and 40°

<u>Jan '00, #32:</u> In the time trials for the 400-meter run at the state sectionals, the 15 runners recorded the times shown in the table below.

400-Meter Run	
Time (sec)	Frequency
50.0–50.9	
51.0–51.9	II
52.0–52.9	⊞I
53.0–53.9	III
54.0–54.9	IIII

a. Using the data from the frequency column, draw a frequency histogram on the grid provided below.

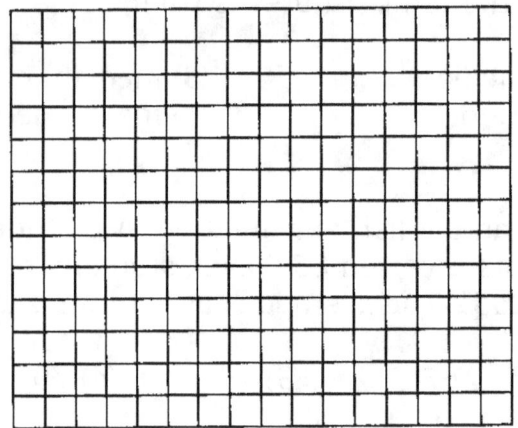

b. What percent of the runners completed the time trial between 52.0 and 53.9 seconds?

<u>Aug '99, #13:</u> If 6 and x have the same mean (average) as 2,4, and 24, what is the value of x?

(1) 5 (2) 10 (3) 14 (4) 36

<u>June '02, #4:</u> During each marking period, there are five tests. If Vanita needs a 65 average to pass this marking period and her first four grades are 60, 72, 55, and 80, what is the lowest score she can earn on the last test to have a passing grade?

(1) 58 (2) 65 (3) 80 (4) 100

<u>June '02, #20:</u> The accompanying diagram is an example of which type of graph?

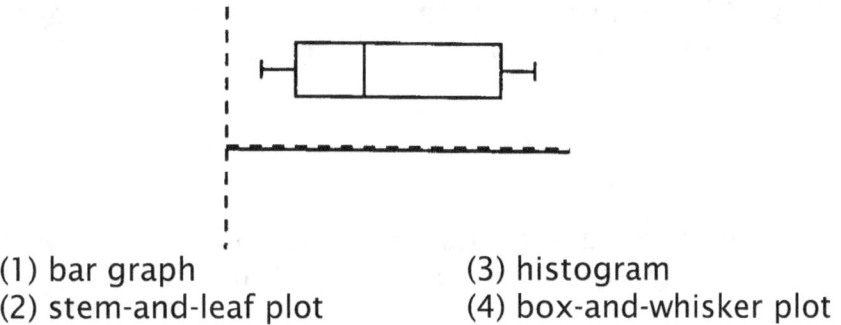

    (1) bar graph   (3) histogram
    (2) stem-and-leaf plot   (4) box-and-whisker plot

<u>Aug '02, #27:</u> Tamika could not remember her scores from five mathematics tests. She did remember that the mean (average) was exactly 80, the median was 81, and the mode was 88. If all her scores were integers with 100 the highest score possible and 0 the lowest score possible, what was the lowest score she could have received on any one test?

<u>Sample #1:</u> For what value of x will 8 and x have the same mean (average) as 27 and 5?

    (1) 1.5   (2) 8   (3) 24   (4) 40

<u>Sample #18:</u> The accompanying histogram shows the scores of students on a Math A test.

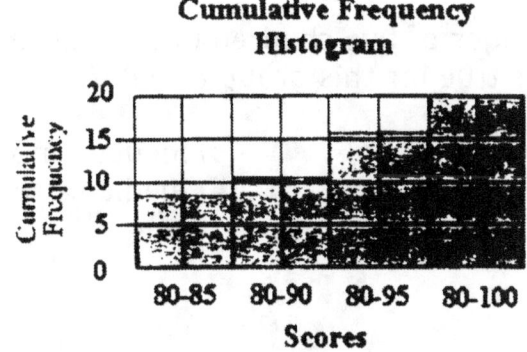

How many students have scores of 96 to 100?

    (1) 55   (2) 20   (3) 5   (4) 4

<u>Sample #26:</u> On his first 5 biology tests, Bob received the following scores: 72, 86, 92, 63, and 77. What test score must Bob earn on his sixth test so that his average (mean score) for all six tests will be 80%?

<u>June '00, #17:</u> For five algebra examinations, Maria has an average of 88. What must she score on the sixth test to bring her average up to exactly 90?

    (1) 92   (2) 94   (3) 98   (4) 100

<u>Jan '03, #1:</u> The accompanying diagram shows a box-and-whisker plot of student test scores on last year's Mathematics A midterm examination.

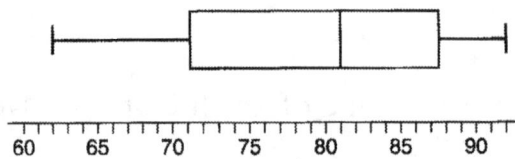

What is the median score?

    (1) 62   (2) 71   (3) 81   (4) 92

<u>Jan '03, #15:</u> The ages of five children in a family are 3,3,5,8, and 19. Which statement is true for this group of data?

    (1) mode>mean   (3) median=mode
    (2) mean>median   (4) median>mean

<u>Jan '03, #34:</u> Sarah's mathematics grades for one marking period were 85, 72, 97, 81, 77, 93, 100, 75, 86, 70, 96, and 80.

    a. Complete the tally sheet and frequency table below, and construct and label a frequency histogram for Sarah's grades using the accompanying grid.

Interval (grades)	Tally	Frequency
61–70		
71–80		
81–90		
91–100		

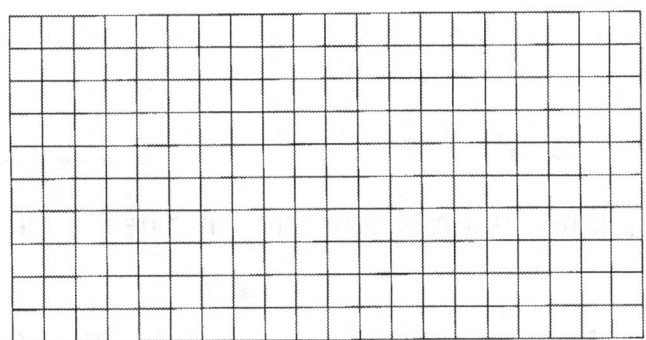

    b. Which interval contains the 75th percentile (upper quartile)?

<u>June '03, #21:</u> The student scores on Mrs. Frederick's mathematics test are shown on the stem-and-leaf plot below.

```
4 | 3

6 | 0  5  5  7  9

7 | 2  5  6  8  9  9  9

9 | 0  1  2  5  9
```

Key: 4 | 3 = 43 points

Find the median of these scores.

Scatter Plots

A scatter plot shows the points of data without connecting them.

Given raw data, it can help you to see what relationship might exist between two variables.

Using the temperature averages for March, plot the days of the month on the
x-axis, and the temperatures on the y-axis using the grid on the next page.

Does there appear to be a trend?_____

Using a straightedge, draw a straight line that nearly matches the trend in the data.

Choose two points on the line and, using the following equations, find the equation of this line.

$$m = \frac{y_2 - y_1}{x_2 - x_1} \qquad\qquad y - y_1 = m(x - x_1)$$

This is called the line of "best fit".

The graphing calculator will also create a scatter plot for us and find the equation of the line of best fit. Now we will only look at the scatter plot, saving the equation for another lesson.

Enter the dates in L_1.
Enter the temperatures in L_2.

Press | 2nd | , | Y= | , | ENTER | .

Turn on Plot 1. Choose the scatter plot type of plot.
Be sure the Xlist: says L_1 and the Ylist: says L_2.

Choose ZoomStat in the Zoom menu.

How does the screen compare with your graph below?_____

_____

Make a sketch of the screen
in the box at the right.

Northern New York Weather Patterns

"Normal" Temperatures and Precipitation for March in Watertown, NY

Source: Intellicast.com

Day of Month	Normal High Temp	Normal Low Temp	Normal Precipitation
1	34.4	15.5	0.1
2	34.8	15.9	0.1
3	35.3	16.4	0.1
4	35.7	16.9	0.1
5	36.1	17.4	0.1
6	36.6	17.9	0.1
7	37	18.4	0.1
8	37.5	18.9	0.1
9	37.9	19.4	0.1
10	38.4	19.8	0.1
11	38.8	20.3	0.1
12	39.3	20.8	0.1
13	39.8	21.2	0.1
14	40.2	21.7	0.1
15	40.6	22.1	0.1
16	41.1	22.6	0.1
17	41.5	23	0.1
18	42	23.4	0.1
19	42.4	23.9	0.1
20	42.8	24.3	0.1
21	43.3	24.7	0.1
22	43.7	25.1	0.1
23	44.1	25.4	0.1
24	44.6	25.8	0.1
25	45	26.2	0.1
26	45.4	26.6	0.1
27	45.8	26.9	0.1
28	46.3	27.3	0.1
29	46.7	27.7	0.1
30	47.1	28	0.1
31	47.5	28.4	0.1

Finding Equations of Lines

The graphing calculator can quickly find the slope and y-intercept of a line using the statistics menu.

We have already used lists in statistics, but now we will see how lists can help us to graph.

Begin by resetting the calculator!

Given the points A(5,2) and B(3,-4) we will enter the points into L_1 and L_2. X's will be in L_1 and Y's will be in L_2.

L_1	L_2	L_3
5	2	
3	-4	

(5,2)
(3,-4)

*** **Be sure that the x's line up with the corresponding y's!!!**

When the lists are complete:

1. Press STAT .

2. Use the right arrow key and choose Calc.

3. Choose 4:LinReg(ax+b).

4. You should now be at the home screen. The calculator is waiting for you to identify the lists you will be using. You also have the option of sending the equation to a Y function in the same step.

a. [2nd] , [1] .

b. [2nd] , [2] .

c. [VARS] ⟶ Y-VARS.

d. Choose 1:Function

e. Choose 1:Y$_1$

f. [ENTER]

Before pressing enter your screen should look like the one on the left; after on the right:

LinReg(ax+b) L$_1$, L$_2$, Y$_1$	LinReg Y=ax+b a=3 b=-13

a is the slope, **b** is the y-intercept.

Press [Y=] .

The first line below the stat plots should now read:

   \ Y$_1$=3X+-13

Press ┌─────────┐ . Sketch the graph in the box below.
      │ GRAPH │
      └─────────┘

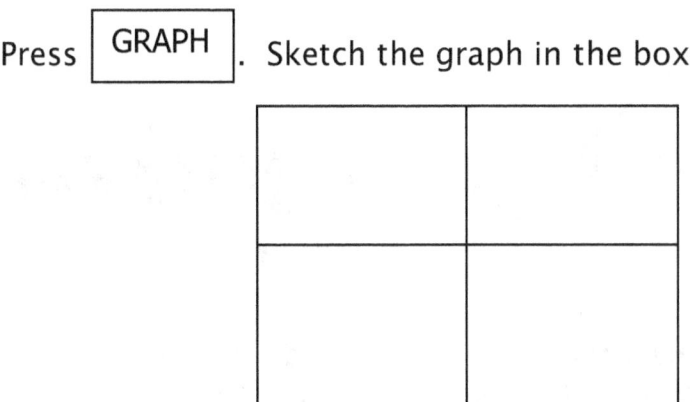

Note: Sending the equation of the line to the Y function and graphing the equation are optional. The slope and y-intercept can be calculated with only the lists entered after the LinReg command. Simply omit steps (c) through (f).

It is also unnecessary to enter the list names if these are your only two lists.

This is a simple illustration of the LinReg command. LinReg is short for Linear Regression. It will prove useful to become familiar with the command with these very short lists. In Math B we will use it for finding the "Line of Best Fit" using longer lists of data.

Practice: Equations of Lines

Using the LinReg (linear regression) command, find the equations of the lines through the following pairs of points and sketch the graph of the line.

1. (1,5), (3,6)

   a=_____

   b=_____

2. (0,5), (-4,8)

   a=_____

   b=_____

3. (0,4), (-1,-1)

   a=_____

   b=_____

4. (2,3), (0,7)

   a=_____

   b=_____

5. (0,0), (9,8)

   a=_____

   b=_____

6. (-1,0), (3,-4)

   a=_____

   b=_____

7. (6,8), (-4,-4)

    a=_____

    b=_____

8. (4,2), (0,-10)

    a=_____

    b=_____

9. (3,3), (2,2)

    a=_____

    b=_____

10. (1,9), (1,7)

    a=_____

    b=_____

11. (1,4), (5,4)

    a=_____

    b=_____

12. (2,7), (2,18)

    a=_____

    b=_____

$$y=ax+b$$

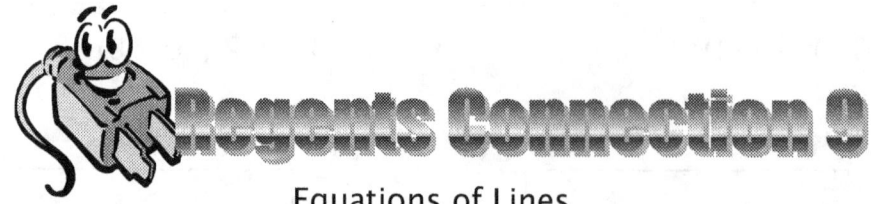

Equations of Lines

<u>Jan '01, #15:</u> What is the slope of line *l* in the accompanying diagram?

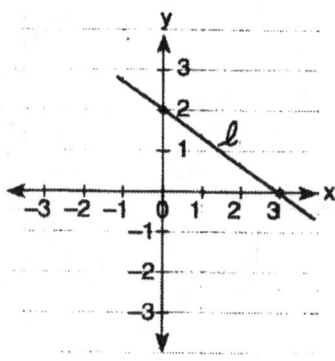

(1) $-\dfrac{3}{2}$ (3) $\dfrac{2}{3}$

(2) $-\dfrac{2}{3}$ (4) $\dfrac{3}{2}$

<u>Aug '99, #19:</u> What is the slope of the line whose equation is $3x-4y-16=0$?

(1) $\dfrac{3}{4}$ (3) 3

(2) $\dfrac{4}{3}$ (4) -4

<u>Aug '99, #29:</u> Line l contains the points (0,4) and (2,0). Show that the point
(-25,81) does or does not lie on line l.

<u>Jun '00, #12:</u> The accompanying figure shows the graph of the equation x=5.

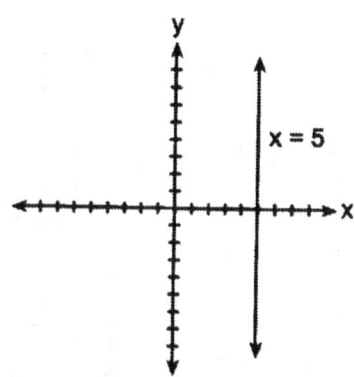

What is the slope of the line x=5?

        (1) 5   (3) 0

        (2) -5   (4) Undefined

<u>Jan '00, #24:</u> A straight line with slope 5 contains the points (1,2) and (3,k). Find the value of k. [The use of the accompanying grid is optional.]

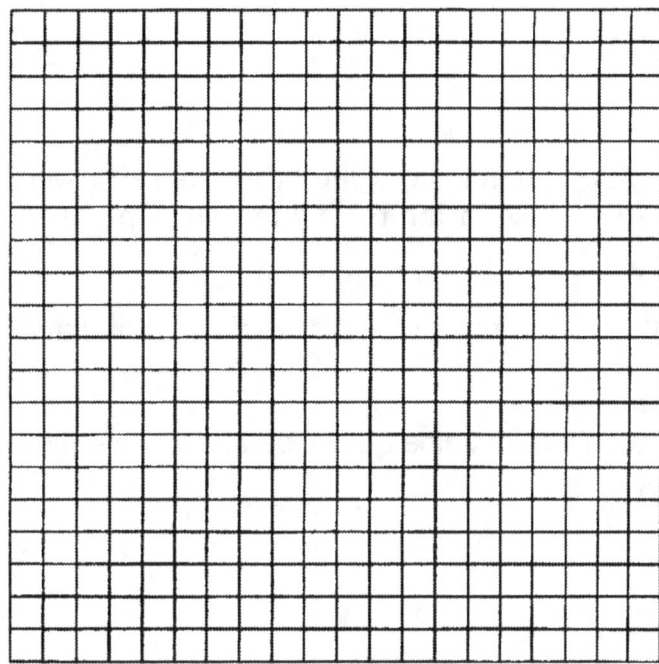

June '99, #18: What is the slope of line l shown in the accompanying diagram?

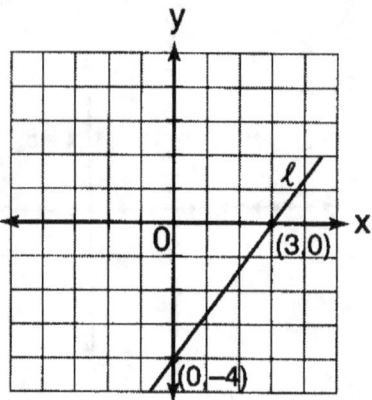

(1) $\dfrac{4}{3}$ (2) $\dfrac{3}{4}$ (3) $-\dfrac{3}{4}$ (4) $-\dfrac{4}{3}$

Aug '00, #9: Which equation represents a line parallel to the line y= 2x - 5?

(1) y= 2x + 5   (3) y= 5x - 2

(2) y= $-\dfrac{1}{2}$x - 5   (4) y= -2x - 5

Aug '01, #30: Shanaya graphed the line represented by the equation y=x-6. Write an equation for a line that is parallel to the given line.

Write an equation for a line that is perpendicular to the given line.

Write an equation for a line that is identical to the given line but has different coefficients.

June '02, #5: What is the slope of the linear equation 5y-10x=-15?

(1) 10   (2) 2   (3) –10   (4) –15

<u>June '02, #10:</u> If two lines are parallel and the slope of one of the lines is m, what is the product of their slopes?

(1) 1   (2) 2m   (3) m²   (4) 0

<u>June '02, #25:</u> Write the equation for the line shown in the accompanying graph. Explain your answer.

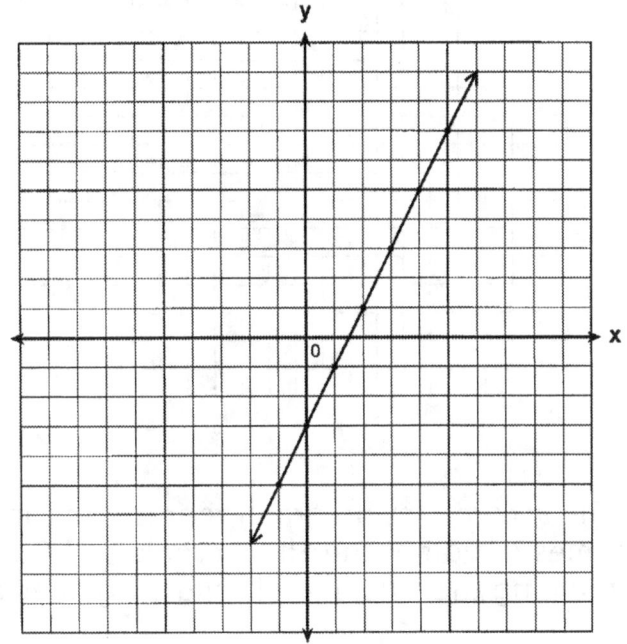

<u>Jan '02, #3:</u> What is the slope of the line whose equation is 2y=5x+4?

(1) 5   (2) 2   (3) $\dfrac{5}{2}$ (4) $\dfrac{2}{5}$

<u>Jan '02, #11:</u> If x and y are defined as indicated by the accompanying table, which equation correctly represents the relationship between x and y?

x	y
2	1
3	3
5	7
7	11

(1) y=x+2   (2) y=2x+2   (3) y=2x+3   (4) y=2x-3

<u>Aug '02, #35:</u> Determine the distance between point A(-1,-3) and point B(5,5). **Write an equation of the perpendicular bisector** of $\overline{AB}$. [The use of the accompanying grid is optional.]

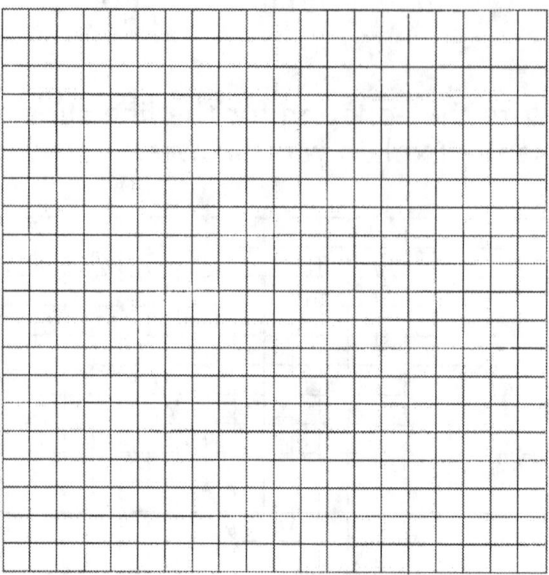

<u>June '00, #25:</u> The accompanying graph represents the yearly cost of playing 0 to 5 games of golf at the Shadybrook Golf Course. What is the total cost of joining the club and playing 10 games during the year?

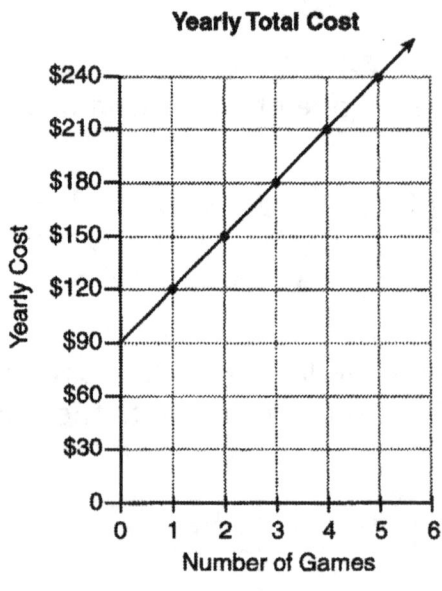

zoom!!

Be sure your calculator is _reset_ before beginning this exercise!

The zoom functions change the way graphs appear on your screen.

We will examine how each of the zoom functions affects the following graphs.

Begin by pressing $\boxed{\text{Y=}}$. Enter the equations below. To move from one equation to the next you can use the down arrow or the ENTER key. To make corrections use the arrow keys to move to the location you want to make the change in.

$$Y_1 = \sqrt{(9-x^2)}$$

$$Y_2 = -\sqrt{(9-x^2)}$$

$$Y_3 = -\sqrt{(4-x^2)}$$

$$Y_4 = \sqrt{((1/2)-(x-1)^2)}+(3/2)$$

$$Y_5 = \sqrt{((1/2)-(x+1)^2)}+(3/2)$$

$$Y_6 = -\sqrt{((1/2)-(x-1)^2)}+(3/2)$$

$$Y_7 = -\sqrt{((1/2)-(x+1)^2)}+(3/2)$$

Be sure you have entered everything _exactly_ as above!

Press GRAPH

Sketch what you see in the box below.

[ ]

Press 2nd ZOOM (FORMAT)

Use the arrow keys to move down and across until AxesOff is highlighted. Press ENTER , then GRAPH

What happened?_____

_____

There are 10 choices in the zoom menu. Some of them will affect the appearance of the current graph and others won't. To see the true effect of each zoom choice press Zoom 6 between each choice. This will take the picture back to its original dimensions. With "ZoomBox" the cursor should be moved, then press ENTER

If nothing happens when using "Zoom In" and "Zoom Out" try pressing

ENTER .

Try each of the choices and sketch your results on the next page.

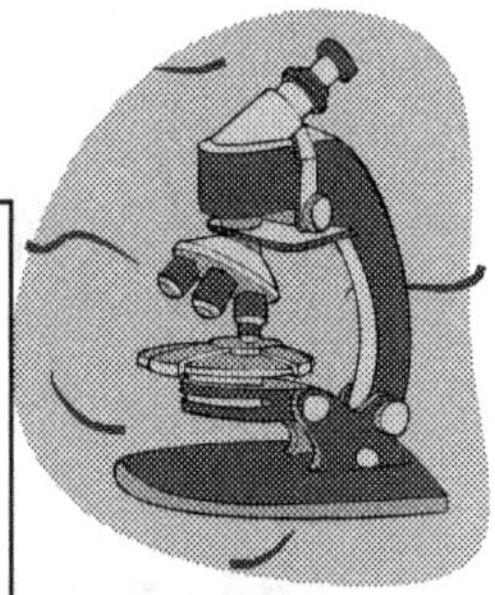

1: ZBox

2: Zoom In

3: Zoom Out

4: ZDecimal

5: ZSquare

6: ZStandard

7: ZTrig

8: ZInteger

9: ZoomStat

10: ZoomFit

Changing Windows

Adjusting the WINDOW settings on the calculator may help you to "see" better.

The standard window shows values from –10 to +10 on both the x and y axes.

Sometimes we need to see beyond these values. Consider the regents question below:

Connor wants to compare Celsius and Farenheit temperatures by drawing a conversion graph. He knows that -40°C=-40°F and that 20°C=68°F. On the accompanying grid, construct the conversion graph and, using the graph, determine the Celsius equivalent of 25°F.

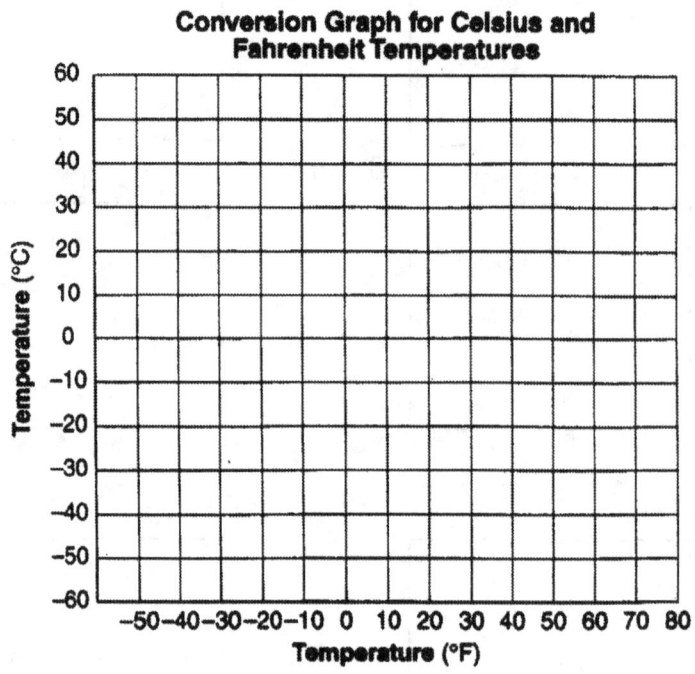

Conversion Graph for Celsius and Fahrenheit Temperatures

[Tip: Drawing in the x and y-axes may help!]

1. Begin by pressing WINDOW

2. The current window settings should be listed on your screen.

3. To change the settings enter the new settings over the previous ones. To decide what the new settings will be, use the grid provided in the question. On the grid, the vertical axis ranges from –60 to +60 and the horizontal axis ranges from –60 to +80 and both increase by tens. Therefore, it is best to use these values for our window settings:

Xmin: -60
Xmax: 80
Xscl: 10
Ymin: -60
Ymax: 60
Yscl: 10

4. With the window correctly set, use the LinReg command with the Fahrenheit temperatures in L_1 and the Celsius temperatures in L_2 and graph in Y_1.

5. Check the graph and draw it carefully on the grid provided.

6. To answer the question we will use the TABLE. The table shows a list of x's and their corresponding y values. To view the table press [2nd] [GRAPH] .

7. Was Fahrenheit the x or y-axis? _____

Begin with the cursor on this column (x).

Use the up and down arrow keys to scroll to 25 in this column.

8. When x is 25, Y_1 is _____ .

Therefore 25°F = _____°C.

How Many Ways Can We Solve a Quadratic Equation?

There are several ways to find the solutions (or roots) of a quadratic equation using the graphing calculator. Your choice may depend on personal preference or how the question is asked.

Let's start with a single quadratic equation and see how many different ways we can arrive at the correct solution(s).

$$x^2-3x+2=0$$

1. First find the solution the "old-fashioned" paper and pencil way.(Factor)

2. If you need to show an algebraic solution you can still use the calculator to help you out.
    If you have factored the equation, but are not sure that your factors are correct, you can use the "=" from the TEST menu to check your answer.

    On the home screen begin with "x^2-3x+2". Now press │ 2nd │
    then │ MATH │ to go to the TEST menu.
    Choose 1:=.

    Then enter the factors you have found. Press │ ENTER │ .
    In this example if you found that the factors were (x-1) and
    (x-2) your screen should look like this:

    ```
    X²-3X+2=(X-1)(X-
    2)
                     1
    ■
    ```

The "1" means that the expression does equal the factors. A "0" would tell you that the factors were incorrect.

CAUTION!! This method is usually reliable, but it can give an incorrect "true" if the currently stored value for x gives an unusual result. To be sure that you are getting the correct result you should store an "unusual" number to x, before you begin. Your age might be a good number to choose.

3. The next method we will investigate is to use EQUATION SOLVER.
   Go to SOLVER and enter x^2-3x+2. Solve for x.

   You now have ONE of the solutions to the equation. Solver can only give you one answer at a time.
   It will give us another solution, but we have to ask for it.

The calculator gives us the solution that was nearest the previous value for x. If we change the beginning value for x to a value larger or smaller than the first solution we should get another solution.

   Try changing the value of x to 10 and solve again.

   Did you get a different solution? _____

   If not try changing the value of x to –10 and solve.
   If you continue to get the same result there may be only one solution, or the second solution may be very unusual and another method may work better.

4. Now we will try graphing the equation.

   Press ☐ Y= ☐

   Enter the equation in Y_1. Your screen should show a parabola . It is not reliable to visually find the roots of the equation, but you can estimate them. If you are having difficulty finding both solutions in SOLVER, this is a quick way to check for a "close" x that you can input.

5. After the equation has been graphed there are two more ways to find the solutions.

The first is to press | 2nd | | TRACE | to find the CALC menu. Choose 2:zero. You will be taken back to the graph screen where you will be asked three questions.

1. Left Bound?
   You should use the left and right arrow keys to position the cursor to the left of one of the points where the parabola meets the x-axis (one of the roots). | ENTER |

2. Right Bound?
   Move the cursor to the right of the same point. | ENTER |

3. Guess?
   Move the cursor as close to the intersection with the x-axis as possible. | ENTER |

You should now see the word "zero" and below this will be the coordinates of the point. In this case it should read x=1 and y=0.

So x=1 is one solution to the equation.

To find the other root you must repeat the process for the other point.

6. One more reliable, but sometimes time consuming (if the roots are not integers) method is to go to the table of values that the calculator created for this function when the equation was entered in the Y= menu. This table is found by pressing

    | 2nd | | GRAPH |

We will explore the uses of the table later. For now just make note that this is a possible way to find the solutions.

7. And, as Math B students know, we could always find the solutions with the quadratic formula.

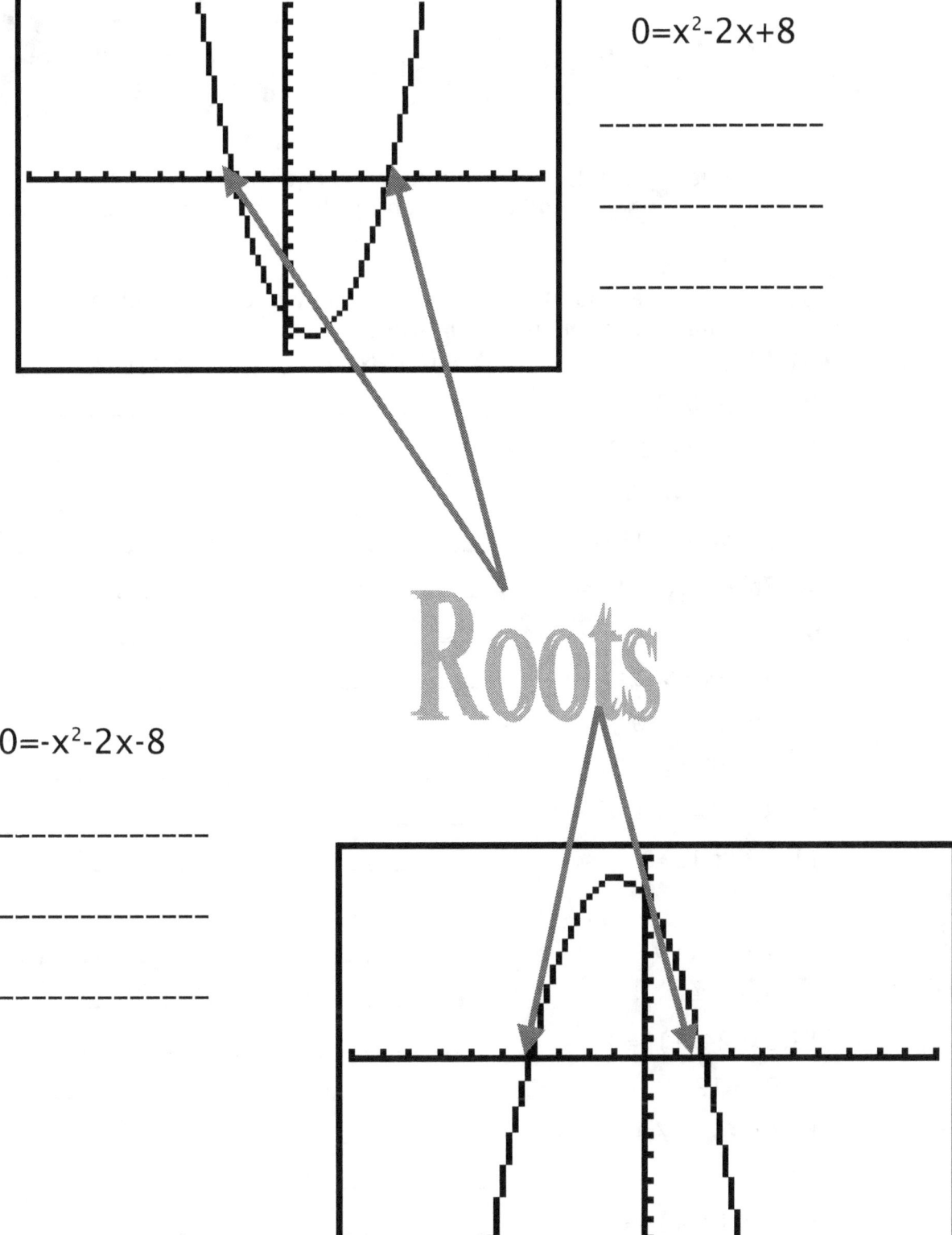

$0 = x^2 - 2x + 8$

_ _ _ _ _ _ _ _ _ _ _ _ _

_ _ _ _ _ _ _ _ _ _ _ _

_ _ _ _ _ _ _ _ _ _ _ _

Roots

$0 = -x^2 - 2x - 8$

_ _ _ _ _ _ _ _ _ _ _ _

_ _ _ _ _ _ _ _ _ _ _ _

_ _ _ _ _ _ _ _ _ _ _ _

Solving Quadratic Equations: Practice

Find the solutions (roots) of each quadratic equation below.
Indicate which method you used for each.

Try to use each method at least once, but only one per problem.
Round decimals to the *nearest hundredth*.

EQUATION:	SOLUTION:	METHOD:
1. $x^2+x-30=0$	--------------	--------------
2. $6x^2-23x+20=0$	--------------	--------------
3. $z^2-14z+40=0$	--------------	--------------
4. $15x^2+34x+15=0$	--------------	--------------
5. $7a^2+22a+3=0$	--------------	--------------
6. $2x^2+x-21=0$	--------------	--------------
7. $4x^2-4x-35=0$	--------------	--------------
8. $2x^2-5x-12=0$	--------------	--------------
9. $6s^2-23s+20=0$	--------------	--------------
10. $3n^2-8n+5=0$	--------------	--------------
11. $2t^2-9t-18=0$	--------------	--------------
12. $9x^2-4=0$	--------------	--------------
13. $100x^2-1=0$	--------------	--------------
14. $8x^2-18=0$	--------------	--------------
15. $9x^2-24x+16=0$	--------------	--------------
16. $3x^2=12x$	--------------	--------------
17. $6c^2+5=-17c$	--------------	--------------
18. $5x-2x^2=0$	--------------	--------------
19. $3x^2-13x+14=0$	--------------	--------------
20. $32x^2-80x+50=0$	--------------	--------------

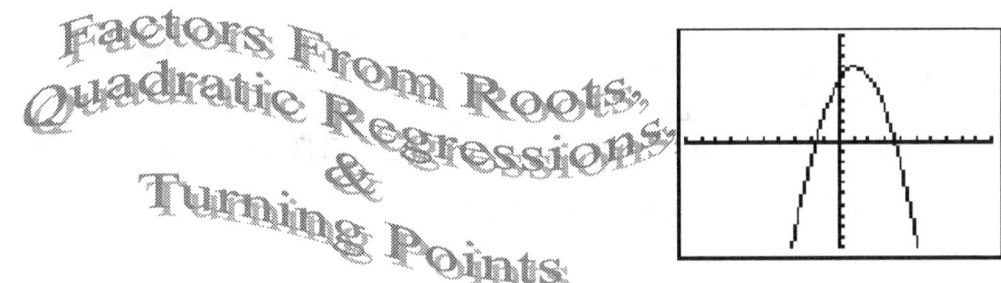

Factors from Roots:

After we have found the solution(s) or roots we can use these to find the factors of a quadratic equation.

Suppose the roots of a quadratic equation are 2 and 4.
     Then, working backwards, we may have obtained these roots from the factors (x-2) and (x-4). Although these are not unique factors for these roots, for Math A and B purposes they are likely. They would be considered the factors of the simplest equation with 2 and 4 as roots.

Practice: Find the factors of the simplest quadratic equation with the given roots.

1. 4, 5 _____   2. –2, 3 _____

3. –1, 7 _____   4. 5, -6 _____

5. –3, -1 _____   6. 0, 2 _____

7. –5, 0 _____   8. –7, -9 _____

Quadratic Regressions:

If you are given the graph of a quadratic equation and can identify at least 3 points, you can use a method very similar to the linear regressions we used to find the equation of a line.

The points (-2,-2), (0,6), and (3,3) are points on the parabola shown in the picture at the top of this page.
   a. Enter the points in L_1 and L_2 just as we did when finding the

     equation of a line, with the x's in L_1 and the corresponding y's in

     L_2.
   b. Press $\boxed{\text{STAT}}$.

c. Go over to CALC.

d. Choose 5:QuadReg.

e. If L_1 and L_2 are your only lists press ENTER. Otherwise, name your x-list and y-list then press ENTER.

f. The screen below should appear.

```
QuadReg
 y=ax²+bx+c
 a=-1
 b=2
 c=6
 R²=1
```

g. Replace the corresponding values in the general form of the quadratic equation to find that the graph is of the equation
$$y=-x^2+2x+6$$

Practice: Use the quadratic regression to find the quadratic equations that satisfy the points given.

1. (-1,12), (0,6), (1,2) _____

2. (-2,-7), (-1,-7), (2,17) _____

3. (-1,1), (0,5), (3,-19) _____

4. (-2,-3), (1,12), (3,32) _____

5. (-1,-7), (0,3), (2,-25) _____

Maximum/Minimum and the Axis of Symmetry:

Every parabola has either a maximum point or a minimum point, also known as its turning point.

With the calculator this point is easy to find.

a. Begin with the equation $y=x^2+7x+6$.
b. Enter this equation in Y_1 and press GRAPH.
c. Decide whether the parabola has a lowest point or a highest point.
d. Press 2nd TRACE.
e. If the parabola had a lowest point choose 3:minimum. If it had a highest point choose 4:maximum.
f. When asked "Left Bound?", move the cursor to a point left of the point you are interested in or input a value you are sure is smaller than the x you are looking for and press ENTER.

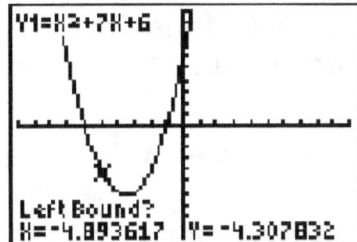

g. When asked "Right Bound?", move the cursor to a point right of the cursor or input a value you are sure is larger than the x you are looking for and press ENTER.

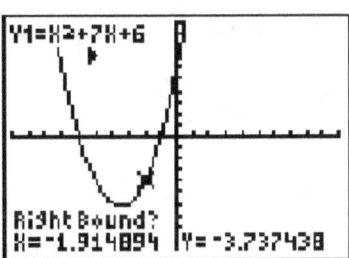

h. Move the cursor near the turning point as your "Guess" or input any value between your left and right bound entries and press ENTER.

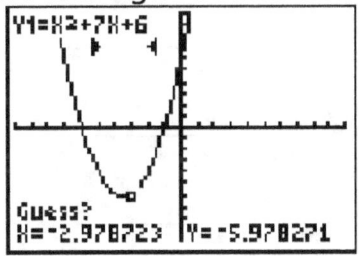

i. The x and y values at the bottom of the screen represent the maximum or minimum (it will tell you which one).

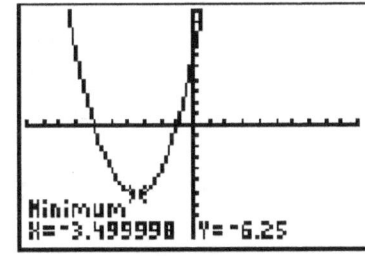

Note that in this example there are several 9's following the decimal value 4. The calculator will occasionally give these trailing 9's or trailing 0's when the actual value is the rounded value (it thought its answer was close enough). The turning point of this parabola is

$$(-3.5, -6.25)$$

j. If you need to identify the axis (or line) of symmetry, notice that the parabola has line symmetry and this line must pass through the turning point. Therefore, if you have found the maximum or minimum, you have found the x-value that this line must pass through. Since it must be a vertical line the equation of the line is

$$x = -3.5$$

Practice: Find the maximum or minimum for each quadratic equation and give the equation of the axis of symmetry for the parabola it represents.

	Turning Point	Axis of Symmetry
1. $y = -x^2 + 5x + 3$	_____	_____
2. $y = 2x^2 + 5x - 9$	_____	_____
3. $y = 3x^2 - 2x - 8$	_____	_____
4. $y = -2x^2 + x + 4$	_____	_____
5. $y = x^2 - 6x + 9$	_____	_____
6. $y = -x^2 + 4x + 12$	_____	_____
7. $y = 4x^2 - 5x - 3$	_____	_____
8. $y = (1/2)x^2 - 2x - 3$	_____	_____
9. $y = -(1/4)x^2 + 3x - 2$	_____	_____
10. $y = (1/3)x^2 + x - 5$	_____	_____

$x = (x-1)+$
$y = (x+1)$

Quadratic Equations

Aug '01, #18: What is the solution set of $m^2-3m-10=0$?

    (1) {5,-2}   (3) {3,-10}
    (2) {2,-5}   (4) {3,10}

Jan '02, #15: What is the solution set of the equation $3x^2 = 48$?

    (1) {-2,-8}   (3) {4,-4}
    (2) {2,8}   (4) {4,4}

Jan '01, #5: One of the factors of $4x^2-9$ is

    (1) (x+3)   (3) (4x-3)
    (2) (2x+3)   (4) (x-3)

Aug '00, #12: The solution set for the equation $x^2-2x-15 = 0$ is

    (1) {5,3}   (3) {-5,3}
    (2) {5,-3}   (4) {-5,-3}

June '01, #4: One root of the equation $2x^2-x-15=0$ is

    (1) $\dfrac{5}{2}$ (3) 3

    (2) $\dfrac{3}{2}$ (4) -3

Aug '99, #26: Solve for x: $x^2+3x-40 = 0$

Jan '02, #33: Javon's homework is to determine the dimensions of his rectangular backyard. He knows that the length is 10 feet more than the width, and the total area is 144 square feet. Write an equation that Javon could use to solve this problem. Then find the dimensions, in feet, of his backyard.

June '99, #9: The larger root of the equation (x+4)(x-3)=0 is

(1) –4   (2) –3   (3) 3   (4) 4

Jan '00, #4: Which expression is a factor of $x^2+2x-15$?

(1) (x-3)   (3) (x+15)
(2) (x+3)   (4) (x-5)

June '01, #9: Factor completely: $3x^2-27$

(1) $3(x-3)^2$ (3) 3(x+3)(x-3)
(2) $3(x^2-27)$ (4) (3x+3)(x-9)

June '01, #31: Find three consecutive odd integers such that the product of the first and the second exceeds the third by 8.

Jan '01, #31: Solve algebraically for x: $\dfrac{1}{x} = \dfrac{x+1}{6}$

Aug '00, #35: Jack is building a rectangular dog pen that he wishes to enclose. The width of the pen is 2 yards less than the length. If the area of the dog pen is 15 square yards, how many yards of fencing would he need to completely enclose the pen?

Aug '01, #3: Written in simplest factored form, the binomial $2x^2-50$ can be expressed as

    (1) 2(x-5)(x-5)   (3) (x-5)(x+5)
    (2) 2(x-5)(x+5)   (4) 2x(x-50)

June '02, #6: Which expression is a factor of $n^2+3n-54$?

    (1) n+6   (3) n-9
    (2) n^2+9 (4) n+9

June '02, #29: Solve for x: $x^2+3x-28=0$

Jan '02, #1: Expressed in factored form, the binominal $4a^2-9b^2$ is equivalent to

    (1) (2a-3b)(2a-3b)   (3) (4a-3b)(a+3b)
    (2) (2a+3b)(2a-3b)   (4) (2a-9b)(2a+b)

Aug '02, #32: A rectangular park is three blocks longer than it is wide. The area of the park is 40 square blocks. If w represents the width, write an equation in terms of w for the area of the park. Find the length and the width of the park.

Sample #6: Which is a factor of $x^2+5x-24$?

    (1) (x+4)   (2) (x-4)   (3) (x+3)   (4) (x-3)

June '00, #35: The area of the rectangular playground enclosure at south School is 500 square meters. The length of the playground is 5 meters longer than the width. Find the dimensions of the playground, in meters. [Only an algebraic solution will be accepted.]

<u>Jan '03, #18:</u> What are the factors of $x^2-10x-24$?

(1) $(x-4)(x+6)$ (2) $(x-4)(x-6)$ (3) $(x-12)(x+2)$ (4) $(x+12)(x-2)$

<u>Jan '03, #26:</u> Three brothers have ages that are consecutive even integers. The product of the first and third boys' ages is 20 more than twice the second boy's age. Find the age of each of the three boys.

<u>Jan '03, #28:</u> The graph of a quadratic equation is shown in the accompanying diagram. The scale on the axes is a unit scale. Write an equation of this graph in standard form.

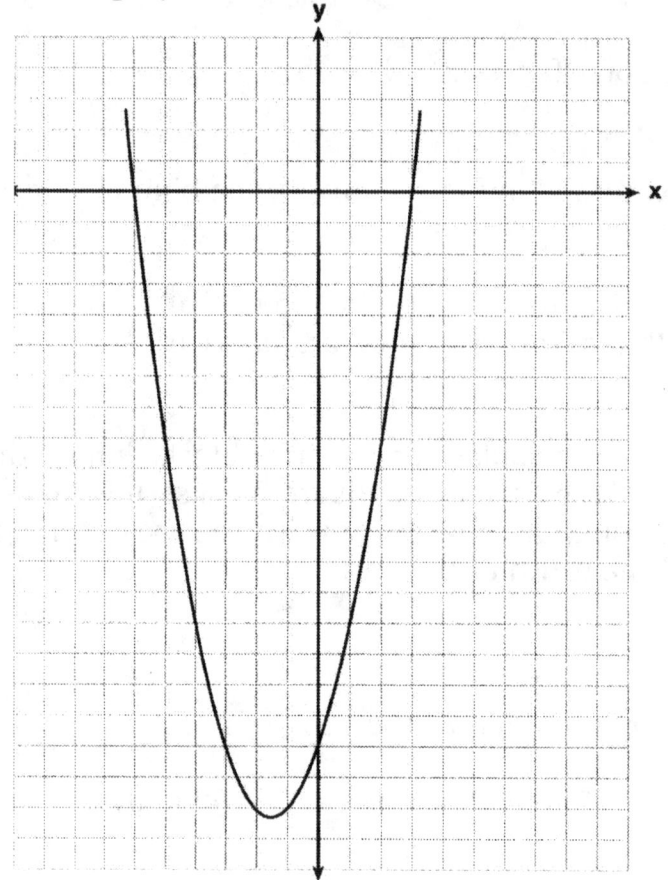

<u>June '03, #13:</u> What is the solution set of the equation
$x^2 - 5x - 24 = 0$?

(1) {-3,8}   (2) {-3,-8}   (3) {3,8}   (4) {3,-8}

Some Quadratic Challenges

1. A ball is thrown straight up at an initial velocity of 54 feet per second. The height of the ball t seconds after it is thrown is given by the formula $h(t)=54t-12t^2$. How many seconds after the ball is thrown will it return to the ground?

   (1) 9.2   (3) 4.5
   (2) 6   (4) 4

2. The solution set of the equation $\sqrt{x+6} = x$ is

   (1) {-2,3}   (3) {3}
   (2) {-2}   (4) { }

3. The roots of the equation $x^2-3x-2=0$ are
   (1) real, rational, and equal
   (2) real, rational, and unequal
   (3) real, irrational, and unequal
   (4) imaginary

If you got all three right you would have received 6 points towards the August 2001 Math B Regents!!

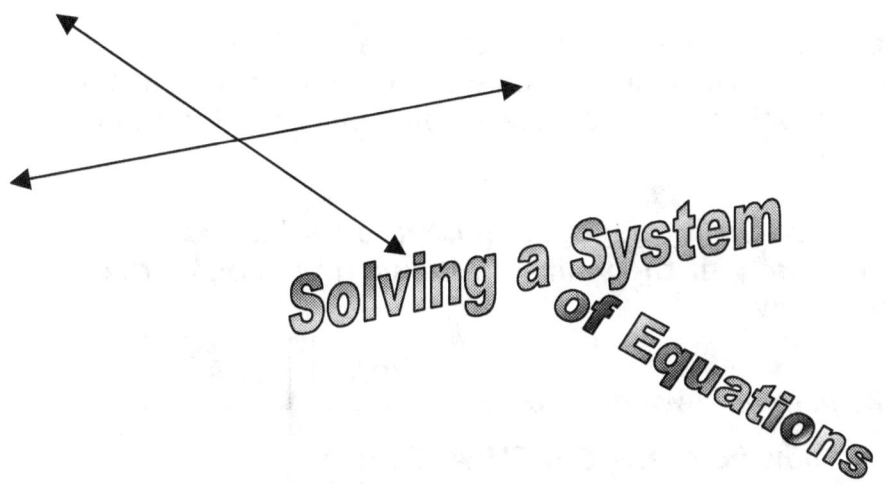

Solving a System of Equations

A system of equations (two lines graphed in the same coordinate plane) can be solved very easily on the graphing calculator.

The first system we will look at is:

$$y=x+6$$
$$y=-2x$$

1. Begin by pressing [Y=] . (Upper left corner.)

2. The cursor should be on Y_1. If there is already an equation there press [CLEAR] .

3. Press [X,T,θ,n] [+] [6] [ENTER]

4. The cursor should now be on Y_2.

5. Press [(-)] [2] [X,T,θ,n] .

6. Press [GRAPH] .

You should see two intersecting lines. The point where these two lines intersect is the "*solution*" of the system.

Do not rely on visually finding the coordinates of the point of intersection. The calculator in standard view gives a distorted image! It also only gives the closest point that the pixels in the screen will allow.

It is best to use the <u>intersect</u> command to find the exact values for x and y at the point of intersection. Follow the directions below:

1. *From the graphing window, press* | 2nd | | TRACE | .

2. *You should now be at the CALCULATE menu.*

3. *Choose 5: Intersect .*

4. *The calculator will ask you to move close to the point of intersection on the first line, then the second one. All you need to do is move the flashing cursor near the point of intersection with the left and right arrow keys and press* | ENTER | *when you are finished.*

5. *It will then ask you to "GUESS?". If you can move the cursor closer from its current location do so. Otherwise just press* | ENTER |

6. *It should now say*

> *Intersection*
> *X=-2 Y=4*

at the bottom of the window.

This means that the point at which the two lines meet is (-2,4).

Use the exercises on the next page to practice.

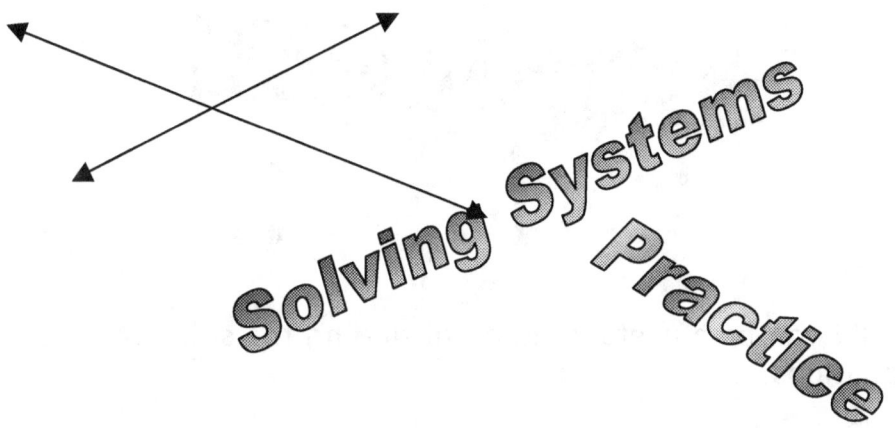

Solving Systems Practice

Find the coordinates of the solution for each of the systems of equations below: (Note that the parentheses have been added for you in this exercise. This will not always be done for you!)

1. y=4x-3
   y=-2x+9

   _____

2. y=(3/2)x-3
   y=x-2

   _____

3. y=-x+4
   y=x-5

   _____

4. y=(-3/2)x+8
   y=(1/4)x+(9/4)

   _____

5. y=-x+6
   y=x

   _____

6. y=2x
   y=2x-(1/2)

   _____

7. y=3x
   y=(3/2)x+9

   _____

8. y=(-3/2)x+(13/2)
   y=(1/4)x+5

   _____

9. y=2x+1
   y=-3x-4

   _____

10. y=-x+10
    y=x-4

    _____

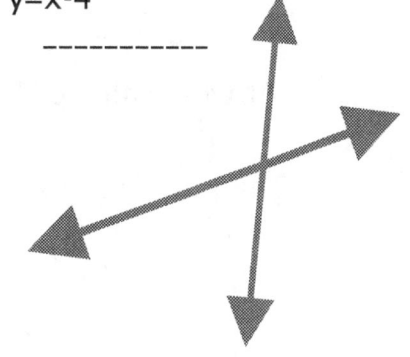

Using the TABLE can be a useful method of finding the solution to a system of equations.

To see what the TABLE can do for us let's enter the equations of two lines:

$Y_1 = 3x - 9$

and

$Y_2 = -2x + 6$

You will still need to enter the equations as if you were going to graph them by pressing [Y=] and entering

the equations in Y_1 and Y_2. Do this NOW.

Next, press [2nd] , [GRAPH] .

Your screen should look like the one below:

X	Y₁	Y₂
0	-9	6
1	-6	4
2	-3	2
3	0	0
4	3	-2
5	6	-4
6	9	-6

X=0

At what value of X is Y_1 equal to Y_2? _____

Then the solution to this system is _____.

Now let's try
$Y_1=4x-12$
and
$Y_2=2x+10$

Notice that when you go to the table there is no common Y value.
As the X's increase the Y values appear to be drawing closer together.
Use the down arrow to scroll to higher X values until you find the X for
which the Y values match.
The solution to the above system of equations is: _____

Note: You can scroll down the list of Y values, but not up. If you need
to scroll up, move the cursor to the X column first.

Examples:

1. Solve: 5x+2y=12 (Note: Solve for y **FIRST**)
              3x-2y=4

                                             (x,y)= _____

2. Solve: -y=13-5x
              y+x=-1

                                             (x,y)= _____

3. Solve: 2x+2y=8
              5x-3y=4

                                             (x,y)= _____

4. Solve: y=5x-10
              y=5x+12

                                             (x,y)= _____

Graphing Calculator for NYS Math A and Beyond

Setting the Table

Try finding the solution to the following system of equations using the TABLE:

$$Y_1=(2/3)X-1$$

$$Y_2=3X+2$$

It appears that the values for Y_1 and Y_2 are close together at X=-1, but they are farther apart at X=0 and X=-2.

This would suggest that the solution must be between 0 and -2 but we cannot see it.
Notice that all of the X values are integers.

Maybe the X in our solution is not an integer.
We can change the X's to decimal values by going to TABLE SET.

Press ⎡ 2nd ⎤ ⎡ WINDOW ⎤ . Your screen should be similar to the one below.

```
TABLE SETUP
 TblStart=█
 ∆Tbl=1
Indpnt: Auto Ask
Depend: Auto Ask
```

Change the TABLE START to the value of X just below where you believe the point of intersection is, in this case -2.
Change ∆Tble to .1 . Go back to the TABLE.
Notice that at X=-1.3 Y_1=-1.867 and Y_2=-1.9. If you need an answer only accurate to the nearest tenth this is close enough.

If you need an answer accurate to the nearest hundredth repeat the above process but change ∆Tble to .01.

Examples: (Use work space to solve for Y if necessary.)

1. $5y=14+3x$

   $3x+8y=12$

   Solution: _____

2. $2x=4y+1$

   $x=y$

   Solution: _____

Practice: Solving Systems with the Table

Solve each of the following systems of equations by using the table of values.

1. y=x-3
   y=-4x+2   Solution: _____

2. y=(1/3)x+12
   y=4x+1   Solution: _____

3. y=2x-3
   y=(3/2)x+1   Solution: _____

4. y=2x-8
   y=x+2   Solution: _____

5. y=10x+1
   y=(1/5)x+1   Solution: _____

6. y=(1/2)x-3
   y=-x+3   Solution: _____

7. y=(1/2)x-6
   y=-4x+3   Solution: _____

Check the Table!

8. y=(2/3)x
    y=(1/8)x-2   Solution: _____
    (Hint: You will need to change ΔTble .)

9. y=(2/7)x
    y=-4x+9   Solution: _____

10. y=(-5/6)x-4
    y=5x+(2/3)   Solution: _____

11. y=3x-2
    y=(1/2)x+1   Solution: _____

12. y=(2/3)x-8
    y=8x+3   Solution: _____

13. y=3x-9
    y=x+6   Solution: _____

14. y=6
    y=(2/3)x-12   Solution: _____
    (Hint: Change table settings to start at 1 and ΔTble=1 .)

June '00, #7: Which ordered pair is the solution of the following system of equations?

$$3x+2y=4$$
$$-2x+2y=24$$

(1) (2,-1)   (3) (-4,8)
(2) (2,-5)   (4) (-4,-8)

Aug '00, #13: What is the value of y in the following system of equations?

$$2x+3y=6$$
$$2x+y=-2$$

(1) 1   (3) -3
(2) 2   (4) 4

Jan '02, #32: When Tony received his weekly allowance, he decided to purchase candy bars for all his friends. Tony bought three Milk Chocolate bars and four Creamy Nougat bars, which cost a total of $4.25 without tax. Then he realized this candy would not be enough for all his friends, so he returned to the store and bought an additional six Milk Chocolate bars and four Creamy Nougat bars, which cost a total of $6.50 without tax. How much did each type of candy bar cost?

June '00, #31: The owner of a movie theater was counting the money from 1 day's ticket sales. He knew that a total of 150 tickets were sold. Adult tickets cost $7.50 each and children's tickets cost $4.75 each. If the total receipts for the day were $891.25, how many of each kind of ticket were sold?

Adult: _____
Children: _____

Jan '01, #34: There were 100 more balcony tickets than main-floor tickets sold for a concert. The balcony tickets sold for $4 and the main-floor tickets sold for $12. The total amount of sales for both types of tickets was $3,056.

a Write an equation or a system of equations that describes the given situation. Define the variables.

b Find the number of balcony tickets that were sold.

June '01, #33: Ramon rented a sprayer and a generator. On his first job, he used each piece of equipment for 6 hours at a total cost of $90. On his second job, he used the sprayer for 4 hours and the generator for 8 hours at a total cost of $100. What was the hourly cost of each piece of equipment?

Jan '00, #27: A truck traveling at a constant rate of 45 miles per hour leaves Albany. One hour later a car traveling at a constant rate of 60 miles per hour also leaves Albany traveling in the same direction on the same highway. How long will it take for the car to catch up to the truck, if both vehicles continue in the same direction on the highway?

Aug '01, #32: The ninth graders at a high school are raising money by selling T-shirts and baseball caps. The number of T-shirts sold was three times the number of caps. The profit they received for each T-shirt sold was $5.00, and the profit on each cap was $2.50. If the students made a total profit of $210, how many T-shirts and how many caps were sold?

Jan '00, #33: A group of 148 people is spending five days at a summer camp. The cook ordered 12 pounds of food for each adult and 9 pounds of food for each child. A total of 1,410 pounds of food was ordered.

    a. Write an equation or a system of equations that describes the above situation and define your variables.

    b. Using your work from part *a*, find:
        (1)   the total number of adults in the group

        (2)   the total number of children in the group

Jan '01, #25: Two trains leave the same station at the same time and travel in opposite directions. One train travels at 80 kilometers per hour and the other at 100 kilometers per hour. In how many hours will they be 900 kilometers apart?

Jan '00, #35: The Excel Cable Company has a monthly fee of $32.00 and an additional charge of $8.00 for each premium channel. The Best Cable Company has a monthly fee of $26.00 and an additional charge of $10.00 for each premium channel. The Horton family is deciding which of these two cable companies to subscribe to.

    a. For what number of premium channels will the total monthly subscription fee for the Excel and Best Cable companies be the same?

    b. The Horton family decides to subscribe to 2 premium channels for a period of one year.
        (1)   Which cable company should they subscribe to in order to spend less money?

        (2)   How much money will the Hortons save in one year by using the less expensive company?

The following is a quadratic-linear system. Solve the same way we have been solving linear systems!

June '00, #18: The graphs of the equations $y=x^2+4x-1$ and $y+3=x$ are drawn on the same set of axes. At which point do the graphs intersect?

        (1) (1,4)   (3) (-2,1)

        (2) (1,-2)   (4) (-2,-5)

June '99, #25: Sara's telephone service costs $21 per month plus $0.25 for each local call, and long-distance calls are extra. Last month, Sara's bill was $36.64, and it included $6.14 in long-distance charges. How many local calls did she make?

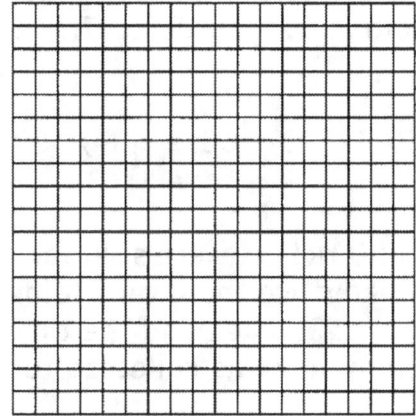

June '99, #35: Solve the following system of equations algebraically or graphically for x and y:

$$y=x^2+2x-1$$
$$y=3x+5$$

For an algebraic solution, show your work here.

For a graphic solution, show your work here.

Jan '00, #22: Mary and Amy had a total of 20 yards of material from which to make costumes. Mary used three times more material to make her costume than Amy used, and 2 yards of material was not used. How many yards of material did Amy use for her costume?

June '01, #10: At a school costume party, seven girls wore masks and nine boys did not. If there were 15 boys at the party and 20 students did not wear masks, what was the total number of students at the party?

    (1) 30   (2) 33   (3) 35   (4) 42

June '01, #17: A hotel charges $20 for the use of its dining room and $2.50 a plate for each dinner. An association gives a dinner and charges $3 a plate but invites four nonpaying guests. If each person has one plate, how many paying persons must attend for the association to collect the exact amount needed to pay the hotel?

    (1) 60   (2) 44   (3) 40   (4) 20

June '01, #23: Ben had twice as many nickels as dimes. Altogether, Ben had $4.20. How many nickels and how many dimes did Ben have?

Aug '99, #16: At a concert, $720 was collected for hot dogs, hamburgers, and soft drinks. All three items sold for $1.00 each. Twice as many hot dogs were sold as hamburgers. Three times as many soft drinks were sold as hamburgers. The number of soft drinks sold was

    (1) 120   (2) 240   (3) 360   (4) 480

Jan '02, #28: A total of 600 tickets were sold for a concert. Twice as many tickets were sold in advance than were sold at the door. If the tickets sold in advance cost $25 each and the tickets sold at the door cost $32 each, how much money was collected for the concert?

Aug '99, #35: Two health clubs offer different membership plans. The graph below represents the total cost of belonging to Club A and Club B for one year.

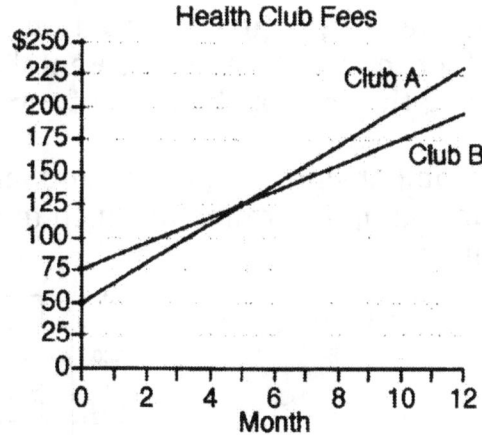

a. If the yearly cost includes a membership fee plus a monthly charge, what is the membership fee for Club A?

b. (1) What is the number of the month when the total cost is the same for both clubs?

(2) What is the total cost for Club A when both plans are the same?

c. What is the monthly charge for Club B?

Jan '01, #30: Juan has a cellular phone that costs $12.95 per month plus $.25 per minute for each call. Tiffany has a cellular phone that costs $14.95 per month plus $.15 per minute for each call. For what number of minutes do the two plans cost the same?

Aug '01, #35: Solve the following system of equations algebraically.

$$y=x^2+4x-2$$
$$y=2x+1$$

June '02, #32: At Ron's Rental, a person can rent a big-screen television for $10 a month plus a one-time "wear-and-tear" fee of $100. At Josie's Rental, the charge is $20 a month and an additional charge of $20 for delivery with no "wear-and-tear" fee.

   a. If c equals the cost, write one equation representing the cost of the rental for m months at Ron's Rental and one equation representing the cost of the rental for m months at Josie's Rental.
   b. On the accompanying grid, graph and label each equation.
   c. From your graph, determine in which month Josie's cost will equal Ron's cost.

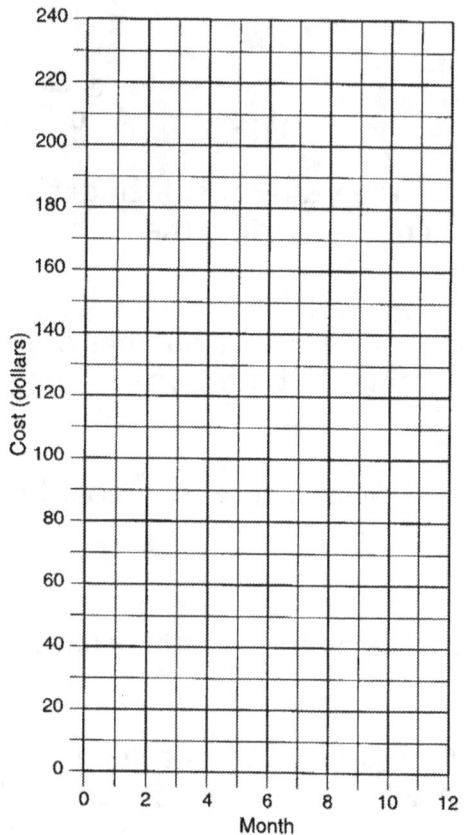

Aug '02, #33: Tanisha and Rachel had lunch at the mall. Tanisha ordered three slices of pizza and two colas. Rachel ordered two slices of pizza and three colas. Tanisha's bill was $6.00, and Rachel's bill was $5.25. What was the price of one slice of pizza? What was the price of one cola?

Sample #31: Two video rental clubs offer two different rental fee plans:

Club A charges $12 for membership and $2 for each rented video.

Club B has a $4 membership fee and charges $4 for each rented video.

The graph below represents the total cost of renting videos from Club A.

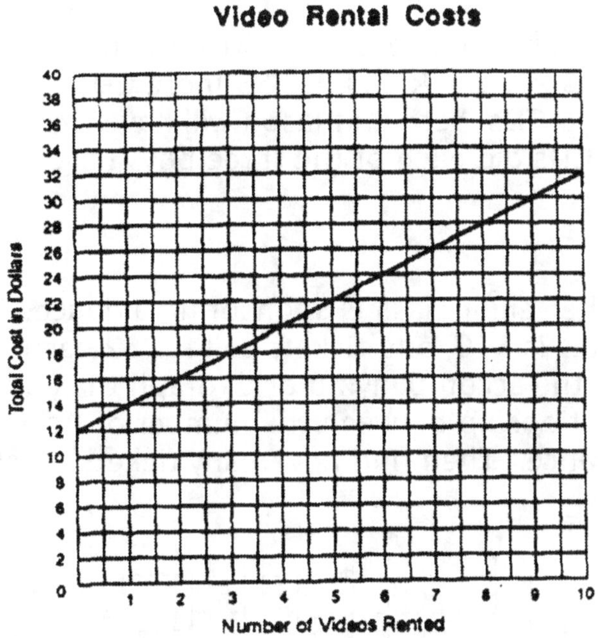

Video Rental Costs

a. On the same set of x,y-axes, draw a line to represent the total cost of renting videos from Club B.
b. For what number of video rentals is it less expensive to belong to Club A? Explain how you arrived at your answer.

Jan '03, #27: Arielle has a collection of grasshoppers and crickets. She has 561 insects in all. The number of grasshoppers is twice the number of crickets. Find the number of each type of insect that she has.

Jan '03, #29: Currently, Tyrone has $60 and his sister has $134. Both get an allowance of $5 each week. Tyrone decides to save his entire allowance, but his sister spends all of hers each week plus an additional $10 each week. After how many weeks will they each have the same amount of money? [Use of a grid is optional.]

<u>Jan '03, #32:</u> Alexandra purchases two doughnuts and three cookies at a doughnut shop and is charged $3.30. Briana purchases five doughnuts and two cookies at the same shop for $4.95. All the doughnuts have the same price and all the cookies have the same price. Find the cost of one doughnut and find the cost of one cookie.

<u>June '03, #26:</u> Seth has one less than twice the number of compact discs (CDs) that Jason has. Raoul has 53 more CDs than Jason has. If Seth gives Jason 25 CDs, Seth and Jason will have the same number of CDs. How many CDs did each of the three boys have to begin with?

<u>June '03, #35:</u> The senior class is sponsoring a dance. The cost of a student disk jockey is $40, and tickets sell for $2 each. Write a linear equation and, on the accompanying grid, graph the equation to represent the relationship between the number of tickets sold and the profit from the dance. Then find how many tickets must be sold to break even.

Math B!

1. A store advertises that during its Labor Day sale $15 will be deducted from every purchase over $100. In addition, after the deduction is taken, the store offers an early-bird discount of 20% to any person who makes a purchase before 10 a.m. If Hakeem makes a purchase of x dollars, x>100, at 8 a.m., what, in terms of x, is the cost of Hakeem's purchase?

   (1) 0.20x-15   (2) 0.02x-3   (3) 0.85x-20   (4) 0.80x-12

2. The cost of a long-distance telephone call is determined by a flat fee for the first 5 minutes and a fixed amount for each additional minute. If a 15-minute telephone call costs $3.25 and a 23-minute call costs $5.17, find the cost of a 30-minute call.

3. The 1999 win-loss statistics for the American League East baseball teams on a particular date is shown in the accompanying chart.

	W	L
New York	52	34
Boston	49	39
Toronto	47	43
Tampa Bay	39	49
Baltimore	36	51

   Find the mean for the number of wins, *w*, and the mean for the number of losses, *l*, and determine if the point (*w,l*) is a point on the line of best fit. Justify your answer.

   w= _____

   l= _____

   Line of Best Fit= _____

4. On a trip, a student drove 40 miles per hour for 2 hours and then drove 30 miles per hour for 3 hours. What is the student's average rate of speed, in miles per hour, for the whole trip?

(1) 34   (2) 35   (3) 36   (4) 37

5. A cellular telephone company has two plans. Plan A charges $11 a month and $0.21 per minute. Plan B charges $20 a month and $0.10 per minute. After how much time, to the nearest minute, will the cost of plan A be equal to the cost of plan B?

(1)   1 hr 22 min   (3) 81 hr 8 min
(2)   1 hr 36 min   (4) 81 hr 48 min

6. The availability of leaded gasoline in New York State is decreasing, as shown in the accompanying table.

Year	1984	1988	1992	1996	2000
Gallons Available in thousands	150	124	104	76	50

Determine a linear relationship for x (years) versus y (gallons available), based on the data given. The data should be entered using the year and gallons available (in thousands), such as (1984,150).

_____

If this relationship continues, determine the number of gallons of leaded gasoline available in New York State in the year 2005.

_____

If this relationship continues, during what year will leaded gasoline first become unavailable in New York State?

_____

7. Island Rent-a-Car charges a car rental fee of $40 plus $5 per hour or fraction of an hour. Wayne's Wheels charges a car rental fee of $25 plus $7.50 per hour or fraction of an hour. Under what conditions does it cost less to rent from Island Rent-a-Car?

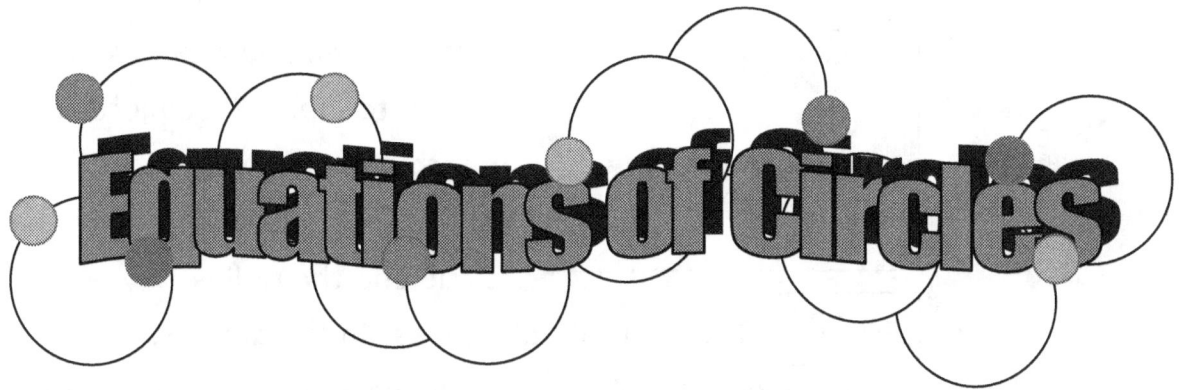

Equations of Circles

A circle in the coordinate plane can be completely described by identifying the center of the circle and how far out the circle reaches from that point (how far it "radiates", hence the term "radius").

The equation of a circle is important to recognize:

$$(x-x_c)^2+(y-y_c)^2=r^2$$

where (x_c,y_c) is the center of the circle and r is the radius.

There are three ways to make a circle on the graphing calculator in the function mode.

1. The equation can be entered in the Y= menu, but it must be done in "half circles". This is how we graphed the "smilie" faces in the ZOOM lesson.

    This can be very cumbersome!

2. An easier way is to DRAW the circle.

    One way to DRAW a circle is:

    a. Begin at the home screen.

    b. Press ⬚2nd⬚ ⬚PRGM⬚ .

    c. Choose9:Circle(

        *** Common mistake: Do not choose 1:ClrDraw!!

        Clr stands for CLEAR not circle!!*********************

d. | Circle(x,y,r) | The circle command requires you to input *three* pieces of information: the x and y values for the center of the circle and the radius.

e. | ENTER | The circle should appear in the graph window.

****Do not press | GRAPH | ! If you have not pressed ENTER on the home screen, your circle will not appear on the graphing screen.

f. Change the ZOOM setting to make the circle a "circle" rather than an "oval". (Hint: ZDecimal or Zsquare work well for this!)

g. Note that sometimes making changes or additions to your screen will cause your circle to "disappear". Just return to the home screen and press 2nd, ENTER to "backspace" to your circle command and re-enter it.

3. The third way to graph a circle also uses the DRAW command, but you will need to begin in the graphing window.

a. Press | GRAPH | .

b. Press | 2nd | | PRGM | .

c. Choose9:Circle(

d. Move the flashing cursor to the point which is to be the center of the circle. Press | ENTER | .

e. Move the cursor up, down, left, or right the number of units the radius of the circle is. Press | ENTER | .

f. The circle should appear.

When finished Clear Draw before beginning another circle. (Choice 1 on the DRAW menu.)

Circle Practice

Part 1. Using the given center and radius, graph the circle using the DRAW command. Sketch the circle in the box provided. Use ZSquare as the zoom setting.

1. (0,0), r=1

2. (1,3), r=3

3. (-2,4), r=5

4. (-3,-5), r=4

5. (0,0), r=10

6. (1,5), r=5

7. (0,8), r=2

8. (-10,0), r=6

9. (0,10), r=4

Part 2. Given the equation of a circle, identify the center and the radius. Then continue as in Part 1.

Example: $(x-2)^2+(y+3)^2=16$

Center=(2,-3)   **Note the sign change!!

Radius=$\sqrt{16}$ =4

1. $(x-1)^2+(y-1)^2=25$
Center=_____
Radius=_____

2. $(x+3)^2+(y-2)^2=81$
Center=_____
Radius=_____

3. $(x-7)^2+(y-1)^2=25$
Center=_____
Radius=_____

4. $(x-2)^2+(y-5)^2=9$
Center=_____
Radius=_____

5. $(x+5)^2+(y+5)^2=16$
Center=_____
Radius=_____

6. $(x-8)^2+(y-7)^2=36$
Center=_____
Radius=_____

7. $x^2+(y+5)^2=64$
Center=_____
Radius=_____

8. $(x+2)^2+y^2=49$
Center=_____
Radius=_____

9. $x^2+y^2=1$
Center=_____
Radius=_____

Fine-Tuning Those Graphs

NYS Guidelines say your graph should contain the following:

1. _____The labeled graph of each equation when more than one function is graphed (no deduction if only one of two is not labeled)

2. _____ Axes appropriately labeled – variables identified

3. _____ Intercepts noted, where appropriate

4. _____ Points of intersection labeled

5. _____ Indicate window used by showing any of the following:
   - a. Intercepts
   - b. Scale on both axes
   - c. Maximum and minimum values of x and y

6. _____ In the graphs of nonlinear functions at least three points should be indicated. Intercepts are acceptable, and when appropriate, the turning point should also be indicated in the graph of the parabola

*** If a student sketches a graph not on a grid for problems where grid use is optional, the above criteria for sketches and graphs still hold.

Examples:

A. Find the solution to the system

$$y=4x-5$$
$$2y=-x+3$$

Screen:

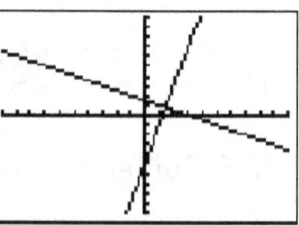

Your graph:

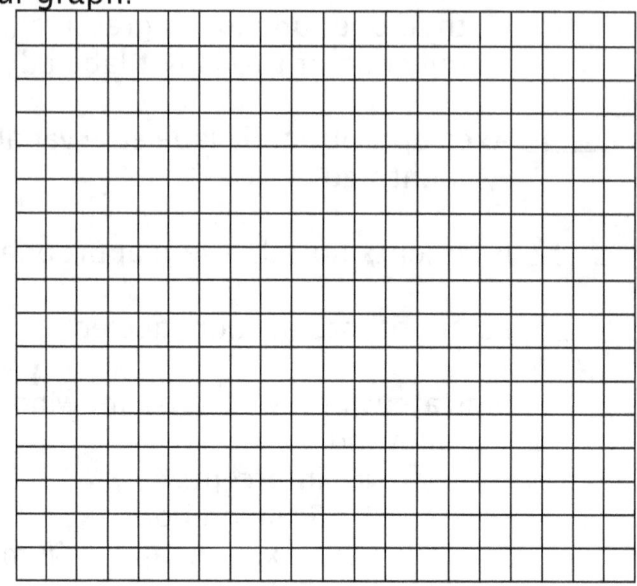

B. When does the graph of $y=x^2-2x-8$ reach its minimum point?

Screen:

Your graph:

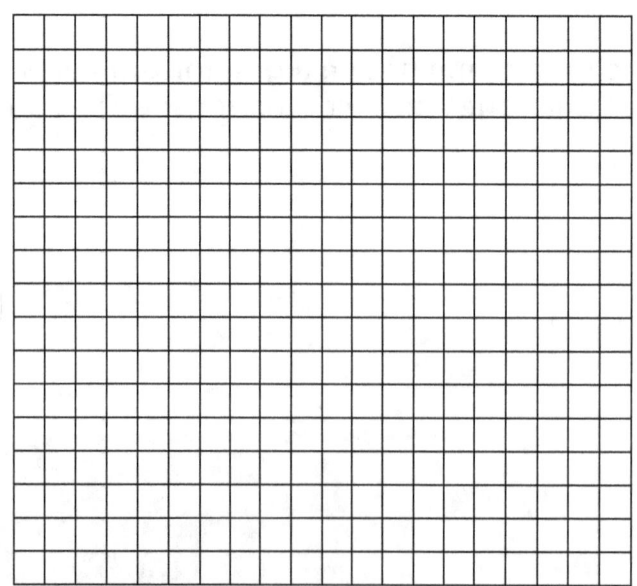

Practice:

1. Graph y=3x-4 and y=-4x+5 on the same axes and find their point of intersection.

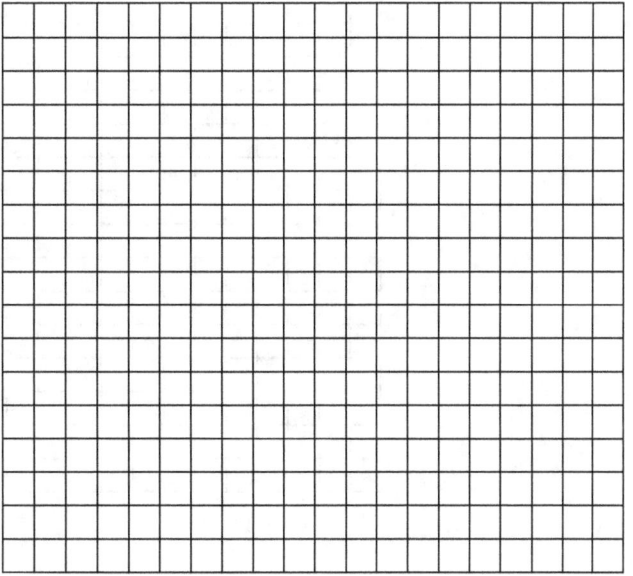

2. Graph y=-2x²+3x+4 and y=2x+3 on the same axes and find their point(s) of intersection.

3. Graphically find how many times the circle $(x-3)^2+(y+4)^2=9$ and the line $y=-(2/3)x+1$ intersect.

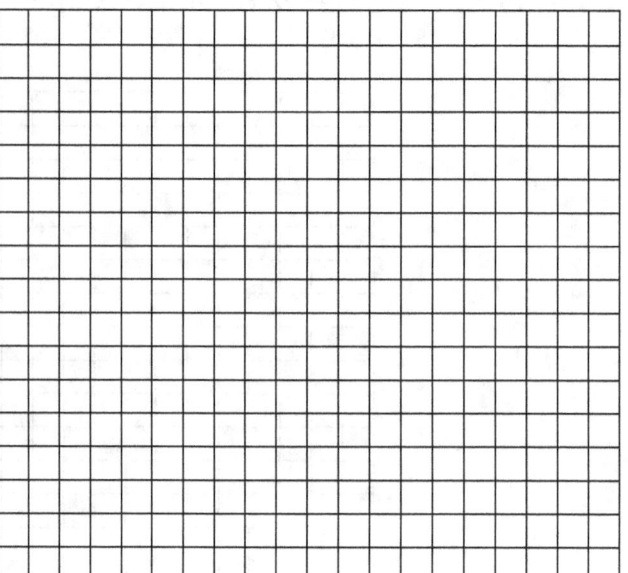

4. Show the solution for the system of inequalities:
$$y<-2x+6$$
$$y\geq 3x^2-2x+2$$

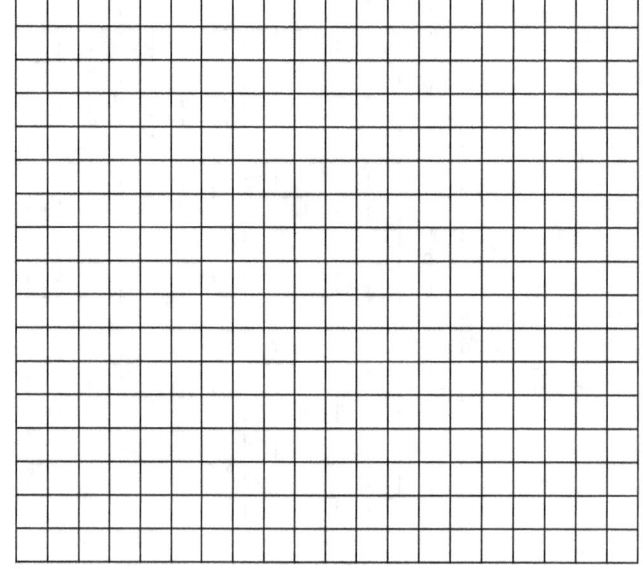

Parabolas and Circles

June '99, #14: What is the diameter of a circle whose circumference is 5?

(1) $\dfrac{2.5}{\pi^2}$ (2) $\dfrac{2.5}{\pi}$ (3) $\dfrac{5}{\pi^2}$ (4) $\dfrac{5}{\pi}$

June '99, #31: A target shown in the accompanying diagram consists of three circles with the same center. The radii of the circles have lengths of 3 inches, 7 inches, and 9 inches.

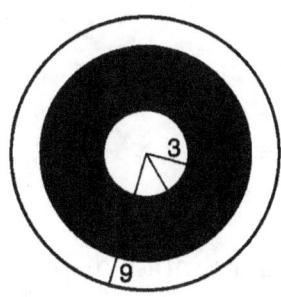

a. What is the area of the shaded region to the nearest tenth of a square inch?

b. To the nearest percent, what percent of the target is shaded?

Jan '00, #12: If the circumference of a circle is 10π inches, what is the area, in square inches, of the circle?

(1) 10π (2) 25π (3) 50π (4) 100π

Aug '00, #17: Which is an equation of the parabola shown in the accompanying diagram?

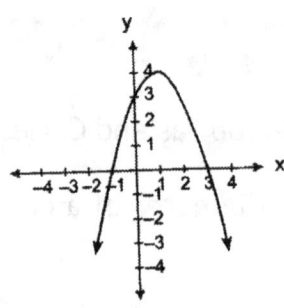

(1) y=-x²+2x+3   (3) y=x²+2x+3
(2) y=-x² – 2x+3   (4) y=x²-2x+3

June '00, #8: Which equation represents a circle whose center is (3,-2)?

(1)   (x+3)²+(y-2)²=4
(2)   (x-3)²+(y+2)²=4
(3)   (x+2)²+(y-3)²=4
(4)   (x-2)²+(y+3)²=4

June '01, #19: What is the total number of points of intersection in the graphs of the equations x²+y²=16 and y=4?

(1) 1   (3) 3
(2) 2   (4) 0

Aug '01, #5: In the accompanying diagram, a circle with radius 4 is inscribed in a square.

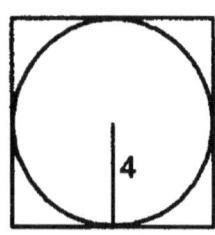

What is the area of the shaded region?

(1) 64-16π   (3) 64π-8π
(2) 16-16π   (4) 16-8π

Jan '00, #29: *a.* On the set of axes provided below, sketch a circle with a radius of 3 and a center at (2,1) and also sketch the graph of the line 2x+y=8.

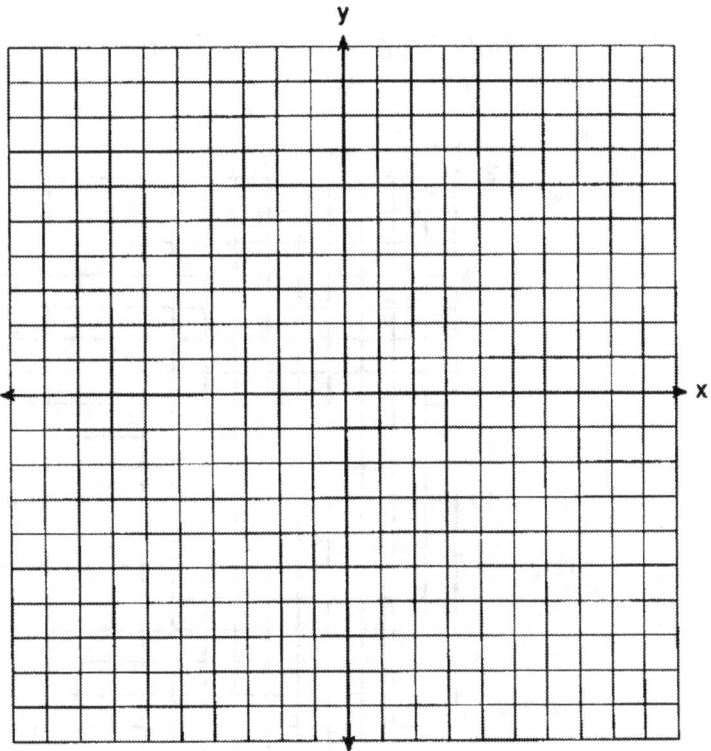

b What is the total number of points of intersection of the two graphs?

Aug '00, #27: To measure the length of a hiking trail, a worker uses a device with a 2-foot-diameter wheel that counts the number of revolutions the wheel makes. If the device reads 1,100.5 revolutions at the end of the trail, how many miles long is the trail, to the nearest tenth of a mile.

<u>Aug '99, #33:</u> An arch is built so that it is 6 feet wide at the base. Its shape can be represented by a parabola with the equation $y=-2x^2+12x$, where y is the height of the arch.

a Graph the parabola from x=0 to x=6 on the grid below.

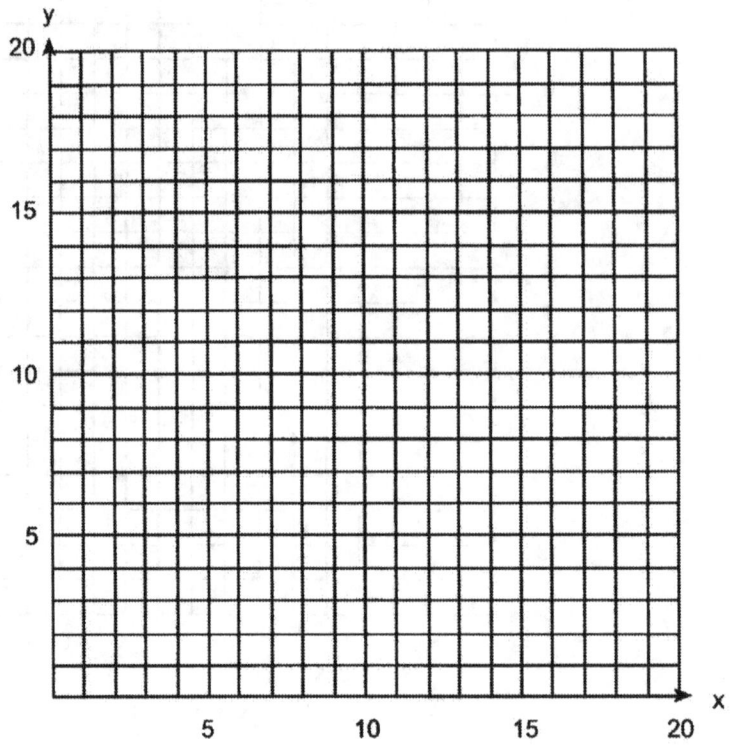

b Determine the maximum height, y, of the arch.

<u>Aug '01, #8:</u> What is the approximate circumference of a circle with radius 3?

(1) 7.07   (2) 9.42   (3) 18.85   (4) 28.27

Jan '00, #31: Amy tossed a ball in the air in such a way that the path of the ball was modeled by the equation $y=-x^2+6x$. In the equation, y represents the height of the ball in feet and x is the time in seconds.

   a Graph $y=-x^2+6x$ for $0 \le x \le 6$ on the grid provided below.

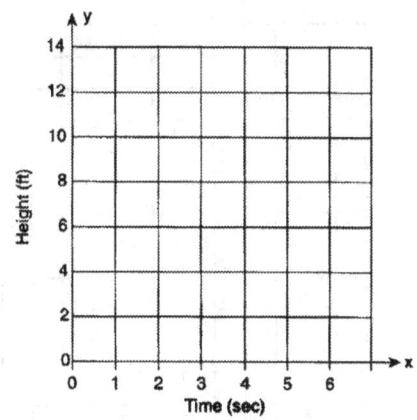

   b At what time, x, is the ball at its highest point?

June '01, #29: Virginia has a circular rug on her square living room floor, as represented in the accompanying diagram. If her entire living room floor measures 100 square feet, what is the area of the part of the floor covered by the rug?

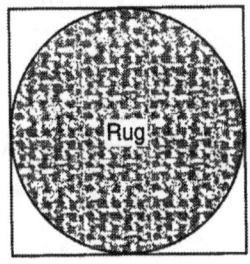

June '02, #15: If the circumference of a circle is doubled, the diameter of the circle

   (1) remains the same   (3) is multiplied by 4
   (2) increases by 2   (4) is doubled

<u>Jan '01, #33:</u> John uses the equation $x^2+y^2=9$ to represent the shape of a garden on graph paper.

    a Graph $x^2+y^2=9$ on the accompanying grid.

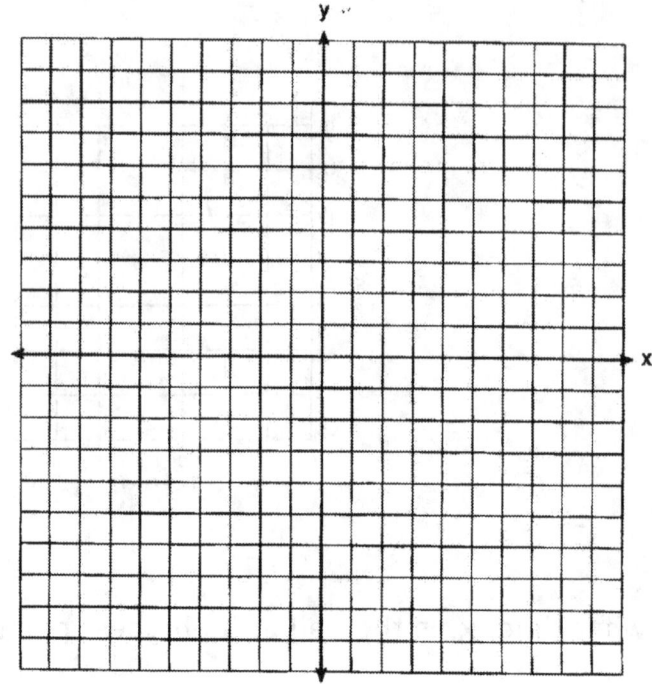

b What is the area of the garden to the *nearest square unit*?

<u>Aug '99, #24:</u> In a recent poll, 600 people were asked whether they liked Chinese food. A circle graph was constructed to show the results. The central angles for two of the three sectors are shown in the accompanying diagram. How many people had no opinion?

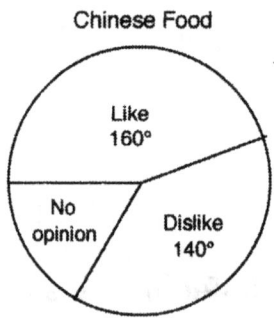

Aug '99, #32: If asphalt pavement costs $0.78 per square foot, determine, to the nearest cent, the cost of paving the shaded circular road with center O, an outside radius of 50 feet, and an inner radius of 36 feet, as shown in the accompanying diagram.

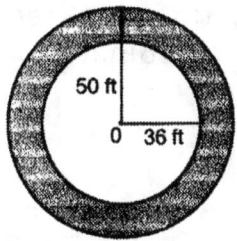

June '02, #28: As shown in the accompanying diagram, radio station KMA is increasing its radio listening radius from 40 miles to 50 miles. How many additional square miles of listening area, to the nearest tenth, will the radio station gain?

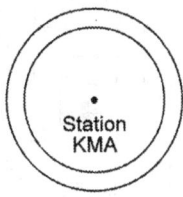

Aug '02, #30: On the accompanying grid, graph a circle whose center is at (0,0) and whose radius is 5. Determine if the point (5,-2) lies on the circle.

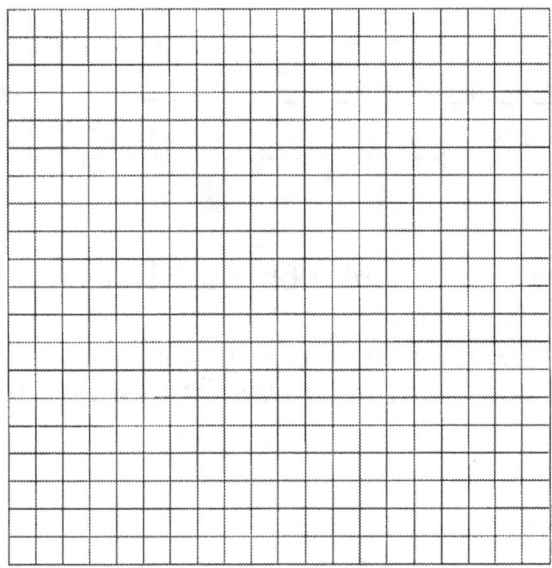

June '02, #35: A rocket is launched from the ground and follows a parabolic path represented by the equation $y=-x^2+10x$. At the same time, a flare is launched from a height of 10 feet and follows a straight path represented by the equation $y=-x+10$. Using the accompanying set of axes, graph the equations that represent the paths of the rocket and the flare, and find the coordinates of the point or points where the paths intersect.

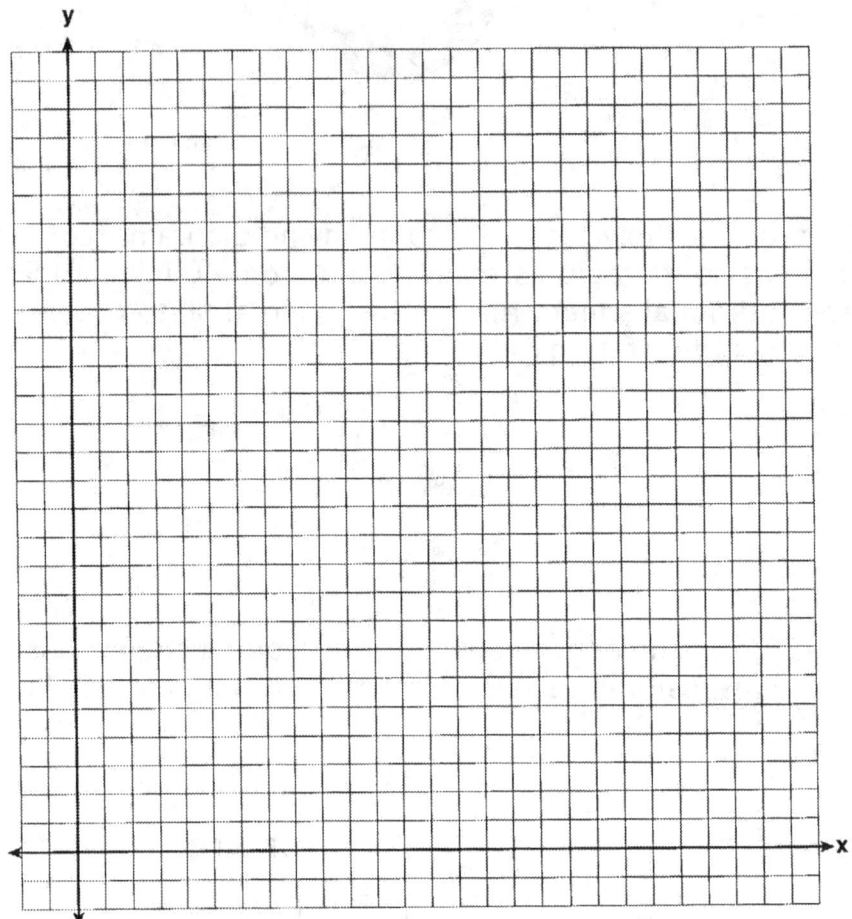

June '00, #28: Tamika has a hard rubber ball whose circumference measures 13 inches. She wants to box it for a gift but can only find cube-shaped boxes of sides 3 inches, 4 inches, 5 inches, or 6 inches. What is the smallest box that the ball will fit into with the top on?

<u>Aug '02, #34:</u> Greg is in a car at the top of a roller-coaster ride. The distance, d, of the car from the ground as the car descends is determined by the equation $d = 144 - 16t^2$, where t is the number of seconds it takes the car to travel down to each point on the ride. How many seconds will it take Greg to reach the ground?

For an algebraic solution show your work here.

For a graphic solution show your work here.

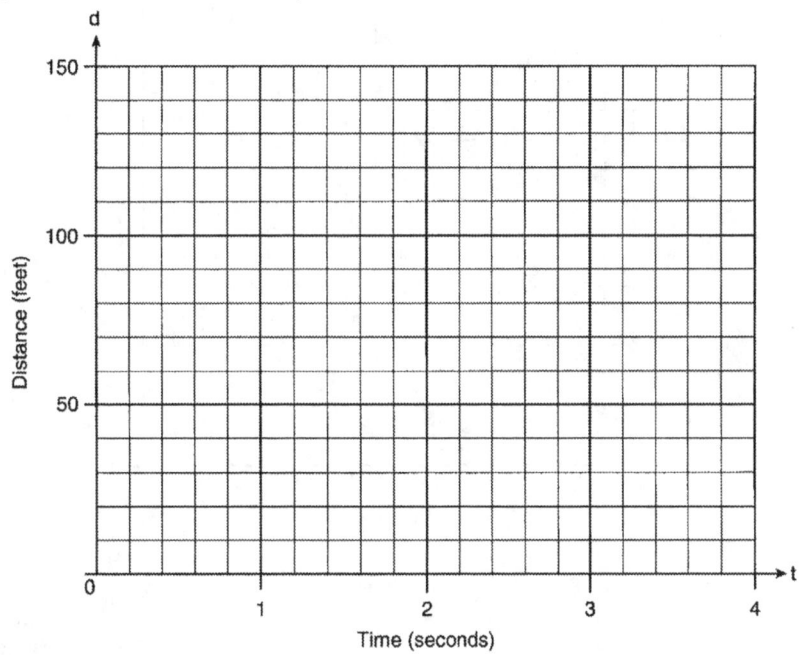

<u>Sample #30:</u> Ms. Brown plans to carpet part of her living room floor. The living room floor is a square 20 feet by 20 feet. She wants to carpet a quarter-circle as shown below.

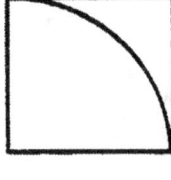

20'

Find, to the nearest square foot, what part of the floor will remain uncarpeted. Show how you arrived at your answer.

June '03, #33: An architect is designing a museum entranceway in the shape of a parabolic arch represented by the equation $y = -x^2 + 20x$, where $0 \le x \le 20$ and all dimensions are expressed in feet. On the accompanying set of axes, sketch a graph of the arch and determine its maximum height, in feet.

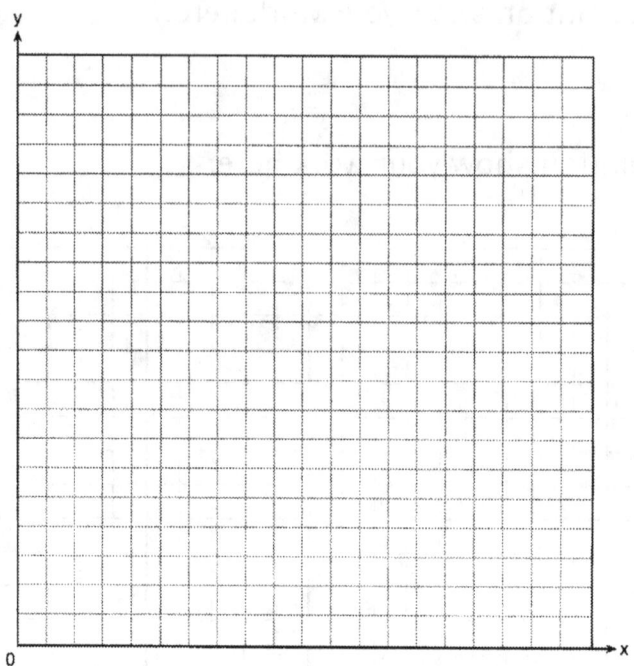

Function ~

Why can't we graph a circle in one step in the Y= menu?

Why can't we graph a vertical line in the Y= menu?

To enter an equation in the Y= menu it must be a *function*.

A *function* is _____
_____
_____
_____

A *relation* is _____
_____
_____
_____

One way to see if the figure graphed is a function or not is to

do a _____ .

On each of the shapes below draw in a vertical line that passes through the shape at two points if it is possible.

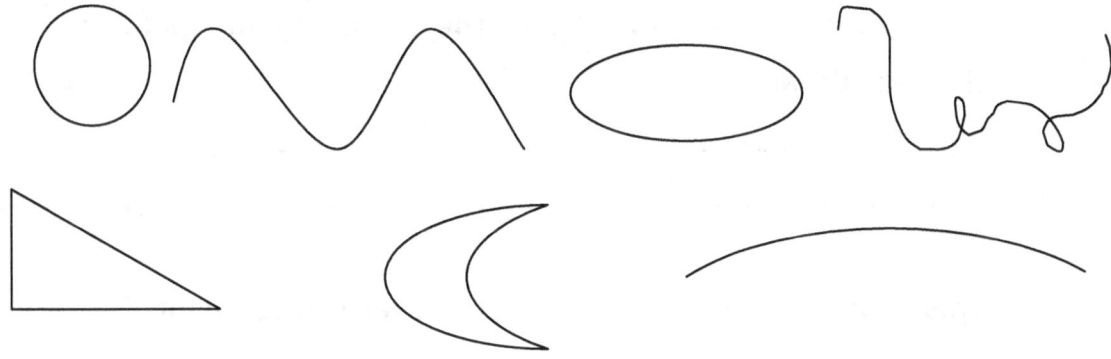

Circle the shapes above that could represent functions. (Hint: Circle the ones you could not draw a vertical line that touched two places on the figure.)

If you cannot graph a function, how can you use the graphing calculator with these shapes?

We have already looked at how to DRAW circles.

We have also seen that vertical lines are impossible to graph in Y=. We can DRAW these too.

Be sure you are beginning in a standard window.

Press [2nd] [PRGM] choose 2:Line(and the screen below should appear.

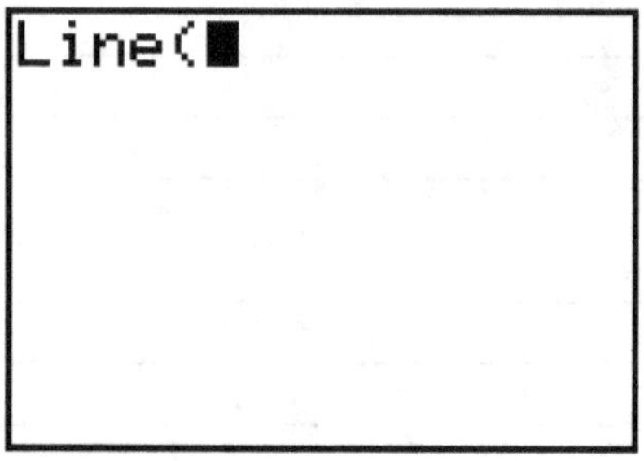

```
Line(■



```

At the prompt input 2,1,2,7) and ENTER.

Sketch the screen that appears:

What are the endpoints of the

segment? _____ _____

What is the relationship between the numbers you input and the line that was drawn?

This demonstrates that we can draw a vertical segment.

Now clear the graph (choice 1 on the DRAW menu).

Press [2nd] [PRGM] and choose 4:Vertical and the screen below should appear.

```
Vertical ■



```

At the prompt input 5 and ENTER.

Sketch the screen that appeared:

Write the equation of the line that

was drawn: _____

How would you make the line x=-2 appear on a graph?

_____

_____

_____

Polygons can be drawn by DRAWing each line segment.

Ex: Graph parallelogram ABCD with vertices A(1,2), B(6,2), C(8,5), D(3,5).
    Enter the lines one at a time. Use 2ⁿᵈ ENTER and replace endpoints to make the job go more quickly.
    The screen on the left shows the home screen after all lines have been drawn. The screen on the right shows the graphing screen as it should look when the parallelogram is complete.

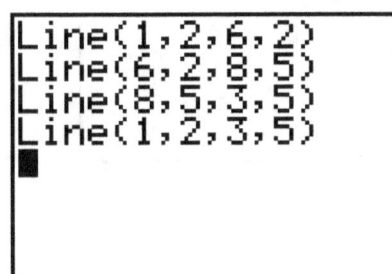
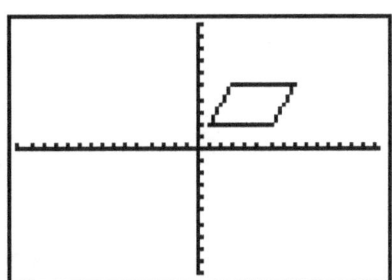

DRAW practice:

DRAW each of the following figures using the Line command. Begin with your screen in Zsquare. Sketch each figure as it appears on your screen. (Hint: Remember that the vertices are listed consecutively and the last must make a segment with the first listed.)

1. DRAW MATH: M(0,-2), A(6,-2), T(6,5), H(0,5)

2. DRAW SUN: S(1,9), U(5,6), N(0,-2)

3. DRAW MAPLE: M(-2,-3), A(-2,6), P(0,9), L(2,6), E(2,-3)

4. DRAW QUAD: Q(-8,-7), U(6,-7), A(8,9), D(-6,9)

5. DRAW THINK: T(-4,7), H(-2,0), I(1,-5), N(4,0), K(6,7)

Math A & B

<u>August B'01, #1:</u> Which relation is not a function?
- (1) $y=2x+4$
- (3) $x=3y-2$
- (2) $y=x^2-4x+3$
- (4) $x=y^2+2x-3$

<u>Jan. B'02, #11:</u> Which relation is a function?
- (1) $x=4$
- (3) $y=\sin(x)$
- (2) $x=y^2+1$
- (4) $x^2+y^2=16$

<u>Jan. B'01: #13:</u> Which equation could represent the relationship between the x and y values shown in the accompanying table?

x	y
0	2
1	3
2	6
3	11
4	18

- (1) $y=x+2$
- (3) $y=x^2$
- (2) $y=x^2+2$
- (4) $y=2^x$

<u>June B'02, #13:</u> Which equation represents a function?
- (1) $4y^2=36-9x^2$
- (3) $x^2+y^2=4$
- (2) $y=x^2-3x-4$
- (4) $x=y^2-6x+8$

August '00, #32: Ashanti is surveying for a new parking lot shaped like a parallelogram. She knows that three of the vertices of parallelogram ABCD are A(0,0), B(5,2), and C(6,5). Find the coordinates of point D and sketch parallelogram ABCD on the accompanying set of axes. Justify mathematically that the figure you have drawn is a parallelogram.

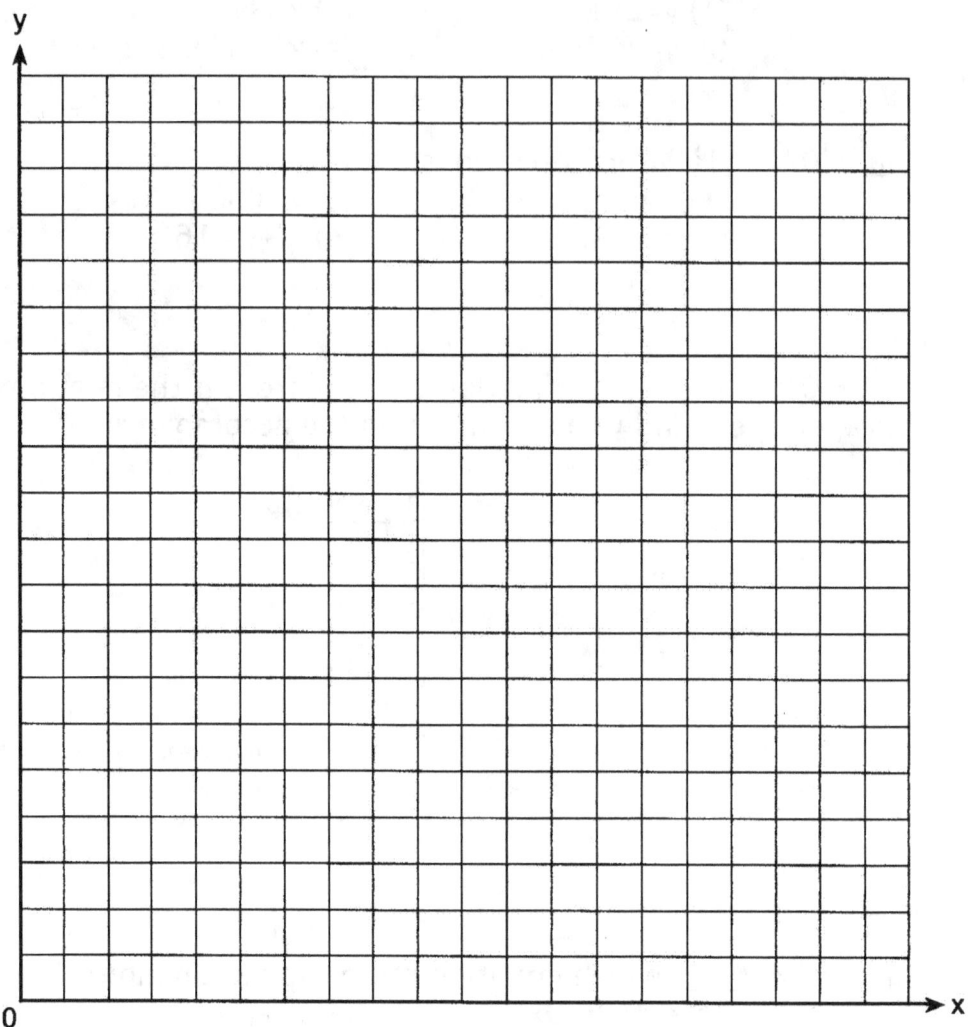

Graphing Inequalities

The TI-83+/TI-84+ will graph inequalities *and* shade them.

Some things you will need to know:

A. The calculator will not tell whether you should be using a dotted or a solid line. You will need to decide which one to use when you copy your graph onto graph paper:

> < or> require a dotted line
> ≥ or ≤ require a solid line

B. You will need to know which general direction needs to be shaded.

C. This method will not work for vertical lines. Ex: x<3

Let's consider the equation   y≤-2x+5 .

1. Press **Y=** .

2. Enter the expression –2x+5 in Y_1.

3. Use the left arrow key to move the cursor to the left until the symbol to the left of Y_1 begins to flash.

4. Pressing enter will change the symbol. These are choices to make the appearance of the line different. We will only be concerned with two of these :

          and

　　　The one on the left appears to point down.
We will choose this one when the less than or less than/equal to symbol is used. (≤ or <)

　　　The one on the right appears to point up.
We will choose this one when the greater than or greater than/equal to symbol is used. (≥ or >).

In the example ≤ is used so press ENTER until 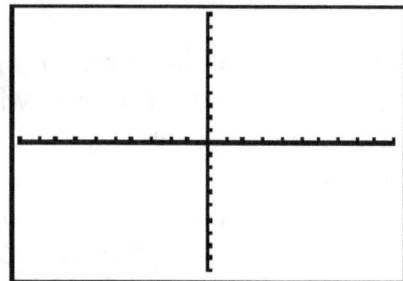 appears beside Y₁.

5. Press GRAPH .

6. Sketch the screen below: Remember that you need to choose whether the line you sketch must be solid or dotted.

Systems of inequalities are also easily done. Follow the same procedure with your first inequality in Y₁ and the second in Y₂.

The calculator will shade them differently making the solution set (the overlap in the shading) clearly seen. It may be helpful to remember that Y₁ is always shaded vertically and Y₂ is always shaded horizontally.

If your screen becomes very light this is normal. This type of graphing requires more energy from the batteries than most work the calculator is asked to do. You can darken the screen by pressing 2ⁿᵈ and holding down the up arrow until you can see it well again. It will return to normal when you leave the graphing screen if no adjustment is made.

With the inequality from the first example still in Y₁, enter y>3x in Y₂. Graph and sketch the screen below:

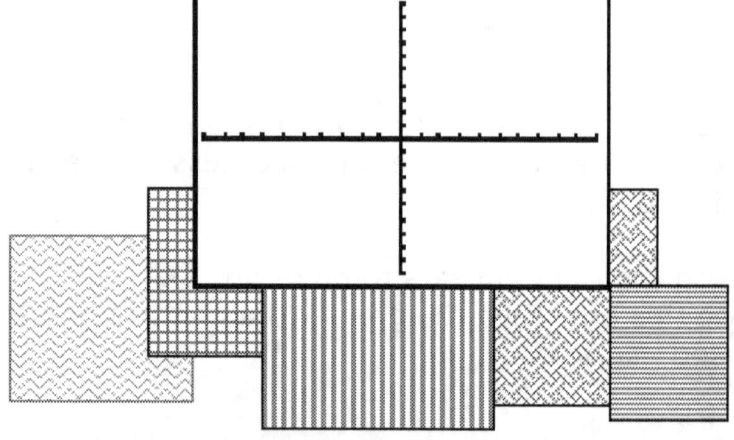

Graphing Inequalities Practice:

Graph and shade each inequality correctly. (Use a standard window.)

1. $y \leq 2x + 3$

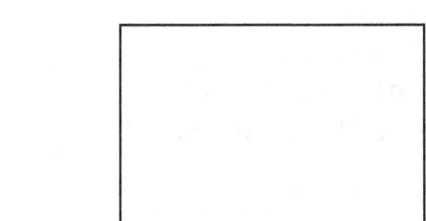

2. $y \geq 2x + 3$

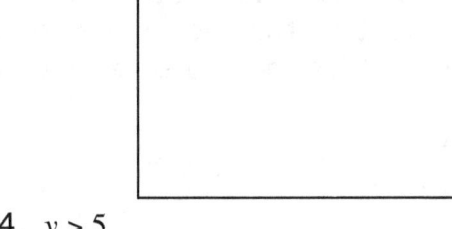

3. $y < 3x$

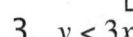

4. $y > 5$

5. $y \leq -3x + 5$

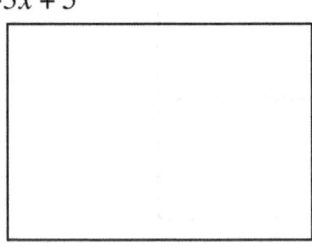

6. $y < -x - 7$

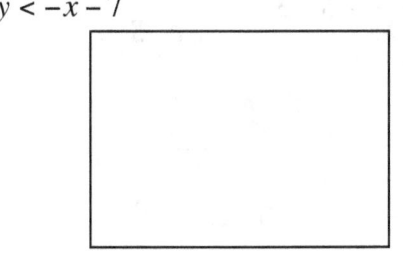

7. $y \geq x - 3$

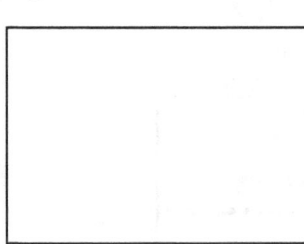

8. $4y > 2x + 8$

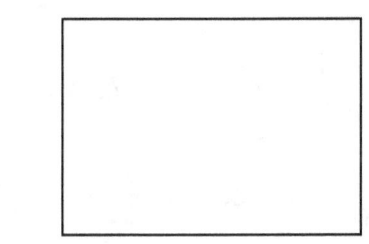

9. $3y < -6x + 9$

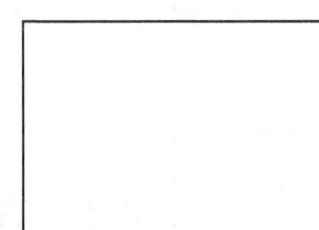

10. $5y \leq 3x + 8$

Practice Graphing *Systems* of Linear Inequalities

For each pair of inequalities
- a. Graph and shade the inequalities on the graphing calculator then sketch the screen in the box given.
- b. Mark the solution set on the sketch with an S.
- c. Give the coordinates of a point **in** the solution set.
- d. Give the coordinates of a point **not** in the solution set.

1. $y \le 3x + 4$
   $y > -2x + 1$

   a. & b.

   c. _____

   d. _____

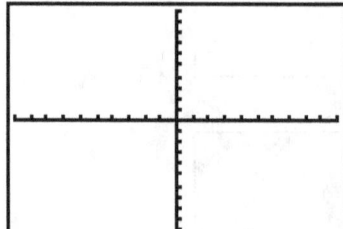

2. $y > -5x - 3$
   $y \le x$

   a. & b.

   c. _____

   d. _____

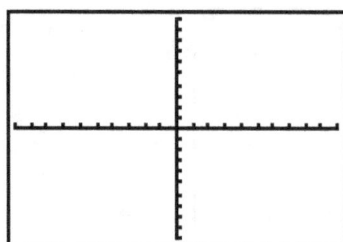

3. $y \le (1/2)x + 1$
   $y < 4x - 3$

   a. & b.

   c. _____

   d. _____

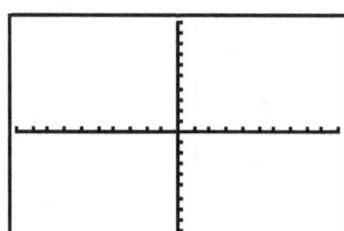

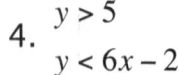

4. $y > 5$
   $y < 6x - 2$

   a. & b.

   c. _____

   d. _____

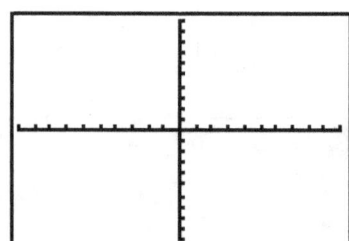

5. $y \geq (2/3)x + 4$
$y > -5x + 4$

       a. & b.

          c. _____

          d._____

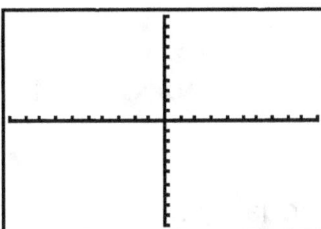

6. $2y < 3x - 7$
$y \geq 4x - 3$

       a. & b.

    Solve for y:

          c. _____

          d._____

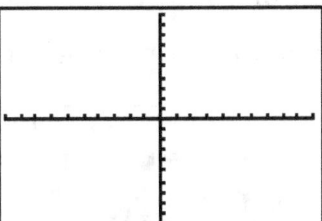

7. $4y > (4/3)x + 8$
$3y \leq x - 3$

       a. & b.

          c. _____

          d._____

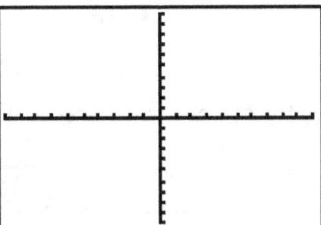

8. $2x - 2y < 6$
$x + y > 1$

       a. & b.

          c. _____

          d._____

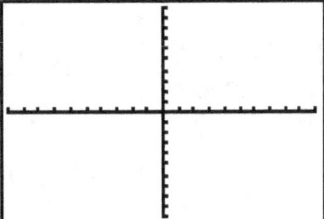

9. $3x < y + 5$
$x \geq 4y - 12$

       a. & b.

          c. _____

          d._____

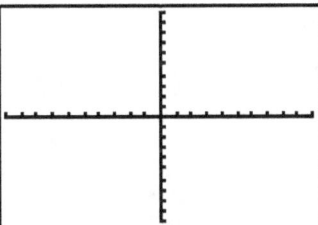

10. $x + 2 > y$
$-x + 2 < y$

       a. & b.

          c. _____

          d._____

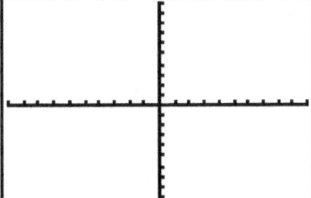

 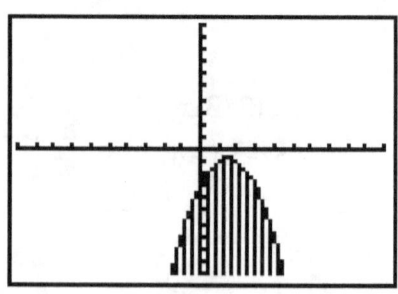

1. Enter the inequality in Y_1 as an equation.

2. Press the left arrow key until the line to the left of Y_1 is flashing.

3. If the inequality is < or ≤ press ENTER until this symbol appears:

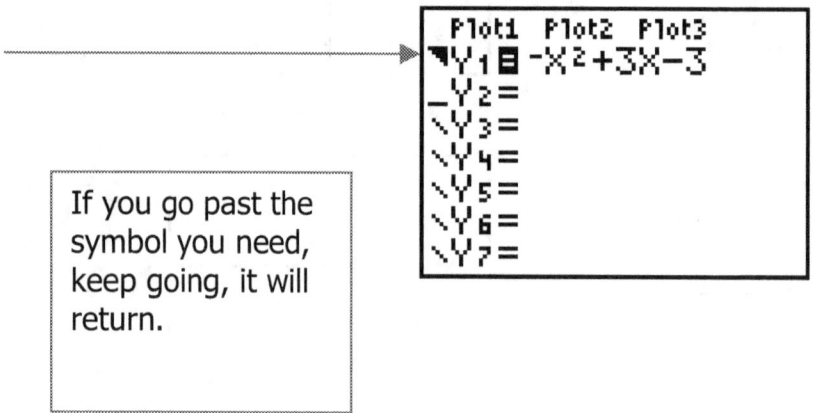

4. If the inequality is > or ≥ press ENTER until this symbol appears:

| Plot1 Plot2 Plot3 |
| ▼Y₁ ⊟ -X²+3X-3 |
| _Y₂= |
| \Y₃= |
| \Y₄= |
| \Y₅= |
| \Y₆= |
| \Y₇= |

If you go past the symbol you need, keep going, it will return.

5. Press GRAPH.

Note that the calculator will not distinguish between < and ≤. You will need to know which one you need.

Practice:
Graph each inequality on the calculator.
Sketch the graph. Use a standard window.

a. y>x²-x-6

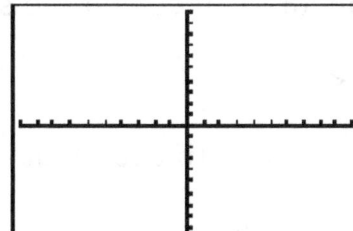

b. y<4x²-9

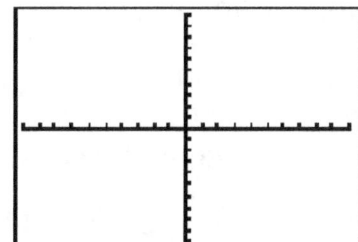

c. y≤x²+2x-15

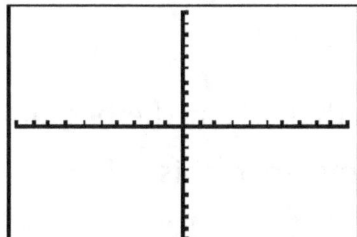

d. y<-2x²+16

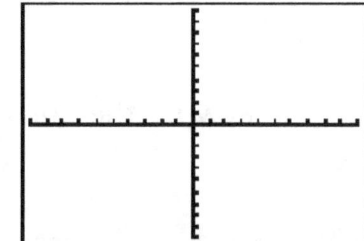

e. y≥-x²+3x-2

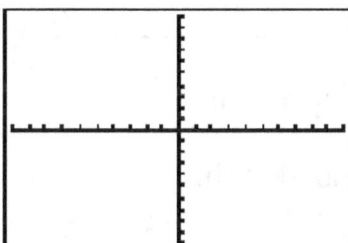

f. y>-4x²+25

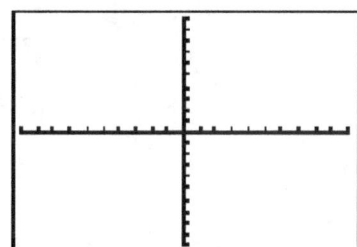

g. y<-(1/4)x² +x+1

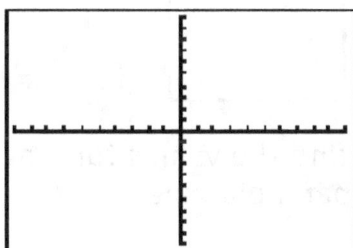

h. y≤(4/9)x²+2x-5

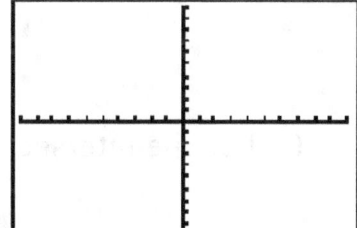

Using Systems to Solve Quadratic Inequalities

Suppose you are given the following quadratic inequality to solve:

$$x^2-2x+1<5$$

1. Begin by separating the inequality into two inequalities.

$$x^2-2x+1< \ldots\ldots\ldots <5$$

2. Place one Y after the first < and another before the second.

$$x^2-2x+1<y \qquad y<5$$

3. Rewrite the one on the left so that y is at the beginning of the inequality. Remember, this will

----

----------------   -----------------

4. Enter these in Y_1 and Y_2.

5. Graph and sketch.

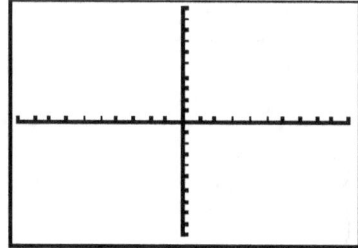

6. Use the intersect command to find the values for x at which the line and the parabola meet.

------------------   -----------------

7. Describe the x values for which both inequalities are true.

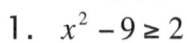

Practice:

a. Enter the inequality as two inequalities in Y_1 and Y_2.

b. Graph in a standard window and sketch the screen neatly.
c. Find the points of intersection.
d. Describe the values of x for which the original inequality are true.

1. $x^2 - 9 \geq 2$ a. & b.

   c. _____ _____

   d. _____

2. $x^2 - 4x + 4 \leq 6$ a. & b.

   c. _____ _____

   d. _____

3. $2x^2 + 3x - 5 < -1$ a. & b.

   c. _____ _____

   d. _____

4. $-x^2 + 25 \leq 4$ a. & b.

   c. _____ _____

   d. _____

5. $-x^2 + 6x - 9 \geq 0$ a. & b.

   c. _____ _____

   d. _____

6. $2x^2 - 10x + 12 > 3$ a. & b.

   c. _____ _____

   d. _____

Mixed Practice
Solving Systems

The following questions require you to find where (for what values of x) the equation/inequality pair is true.

    a. Begin by graphing and shading the pair and sketching the graph in the box provided.
    b. Find all points of intersection.
    c. Determine which x-values will make the statement true.
    d. Graph these values on the number line.

Example: $y = 3x + 3$ and $y \leq -x^2 - 2x + 3$

    a.

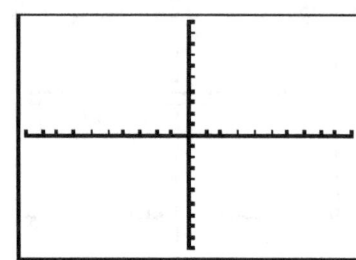

    b. _____ _____

    c. _____

    d. ⟵————————————⟶

1. $y = -x + 2$ and $y > 2x^2 + x - 4$

    a.

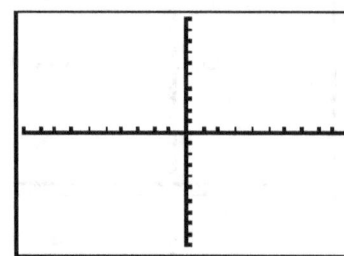

    b. _____ _____

    c. _____

    d. ⟵————————————⟶

2. $y \geq 2x - 1$ and $y = 3x^2 + x - 4$

    a.

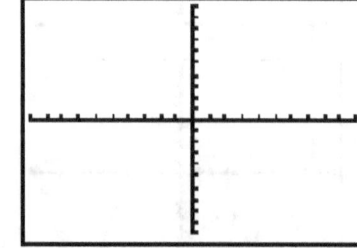

    b. _____ _____

    c. _____

    d. ⟵————————————⟶

3. $y < 3x$ and $y = 2x^2 - 2x - 3$

a.
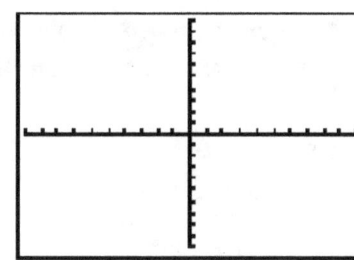

b. _____ _____

c. _____

d. $\longleftrightarrow$

4. $y > (1/3)x - 2$ and $y = x^2 - 4x - 5$

a.
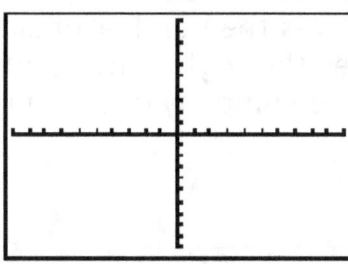

b. _____ _____

c. _____

d. $\longleftrightarrow$

5. $y \leq (2/5)x + 1$ and $y = (1/4)x^2 - 2x + 3$

a.
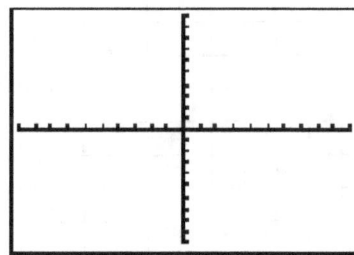

b. _____ _____

c. _____

d. $\longleftrightarrow$

6. $y = 1$ and $y > x^2 + 2x - 8$

a.
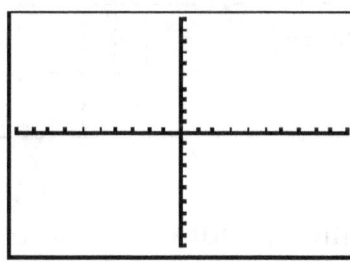

b. _____ _____

c. _____

d. $\longleftrightarrow$

Inequalities

Jan '02, #34: A company manufactures bicycles and skateboards. The company's daily production of bicycles cannot exceed 10, and its daily production of skateboards must be less than or equal to 12. The combined number of bicycles and skateboards cannot be more than 16. If x is the number of bicycles and y is the number of skateboards, graph on the accompanying set of axes the region that contains the number of bicycles and skateboards the company can manufacture daily.

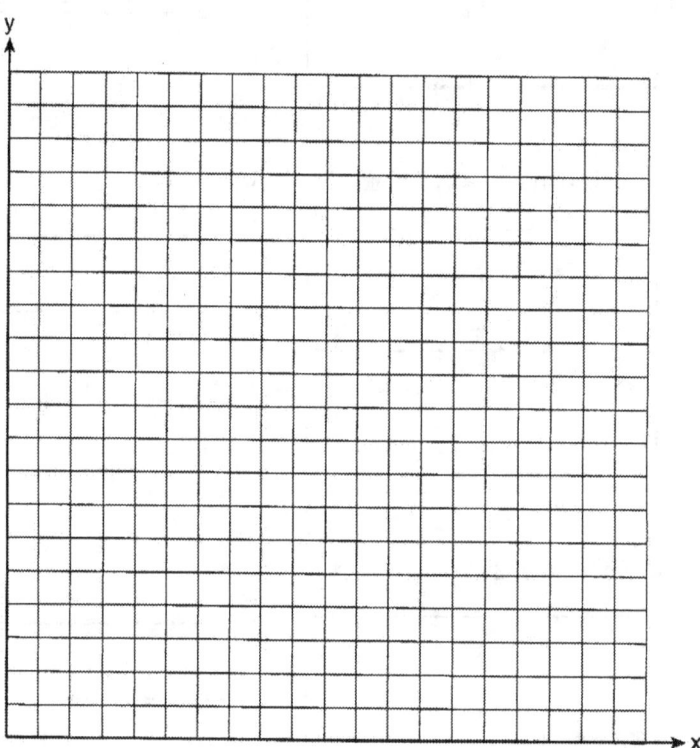

Aug '02, #20: In the graph of y≤-x, which quadrant is completely shaded?

(1) I   (3) III

(2) II   (4) IV

<u>Aug '02, #24:</u> A doughnut shop charges $0.70 for each doughnut and $0.30 for a carryout box. Shirley has $5.00 to spend. At most, how many doughnuts can she buy if she also wants them in one carryout box?

Challenge:
<u>Aug B'02, #3:</u> What is the solution of the inequality $|x+3| \leq 5$?

    (1) $-8 \leq x \leq 2$ (3) $x \leq -8$ or $x \geq 2$
    (2) $-2 \leq x \leq 8$ (4) $x \leq -2$ or $x \geq 8$

<u>June B'01, #28:</u> A homeowner wants to increase the size of a rectangular deck that now measures 15 feet by 20 feet, but building code laws state that a homeowner cannot have a deck larger than 900 square feet. If the length and the width are to be increased by the same amount, find, to the nearest tenth, the maximum number of feet that the length of the deck may be increased in size legally.

<u>Aug. B'01, #2:</u> The solution set of $|3x+2| < 1$ contains
    (1)   only negative real numbers
    (2)   only positive real numbers
    (3)   both positive and negative real numbers
    (4)   no real numbers

The Story of SOH CAH TOA
Source: The Math Forum Ask Dr. Math http://mathforum.org/dr.math/

A young brave, frustrated by his inability to understand the geometric constructions of his tribe's battle dress, kicked out in anger against a stone and crushed his big toe. Fortunately, he learned from this experience, and began to use study and concentration to solve his problems rather than violence. This was especially effective in his study of math, and he went on to become the wisest man of his tribe.

He studied many aspects of trigonometry; and even today we remember many of the functions by his name.

When he became an adult, the tribal priest gave him a name that reflected his special nature – one that reminded them of his great discoveries and of the event that changed his life. Because he was troubled throughout his life by the problematic foot, he was constantly at the edge of the river, soaking his aches in the cooling waters.

For that behavior, he was named Chief *Soh Cah Toa*.

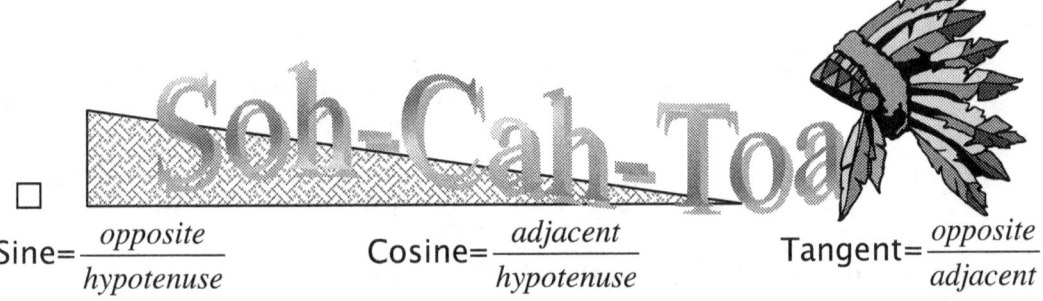

Soh-Cah-Toa

Sine=$\dfrac{opposite}{hypotenuse}$ Cosine=$\dfrac{adjacent}{hypotenuse}$ Tangent=$\dfrac{opposite}{adjacent}$

When the TI-83+/84+ has been reset it will be in _____ mode.

To solve the trigonometry ratios above you will need to change the mode to_____ .

Press | MODE | . Press the down arrow twice and the right arrow once so that Degree is highlighted. Press | ENTER | . Then CLEAR or QUIT to go back to the home   screen.

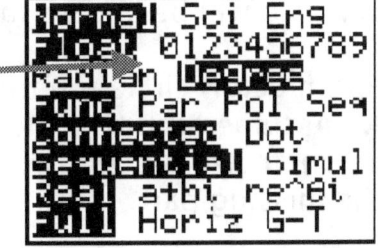

The ratios can be done on the home screen if you are familiar with them or they can be done in SOLVER:

(Use the SIN, COS, and TAN keys.)
Sin (x)-O/H
Cos (x)-A/H
Tan (x)-O/A

Be sure to use the parentheses correctly. The calculator automatically enters the beginning of the set of parentheses when you choose the trigonometry function. You must end them after the part of the equation that represents the degrees.

Also sketch a diagram that represents the problem first so that you are using the correct function.

Hints:
An angle of elevation is an angle that begins horizontally and moves up.

An angle of depression or of descent begins horizontally and moves down.

Practice Identifying the Ratio to Use:

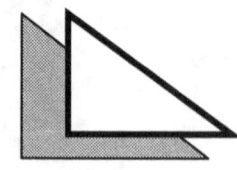

If you use a mnemonic device to remember the trigonometry ratios write it here:

_____

_____

The adjacent side is: _____

The opposite side is: _____

The hypotenuse is: _____

The angle is measured in: _____

In each right triangle trigonometry problem, you will be given two sides and an angle. Two will be known values, the third you will need to solve for.

    a. In each diagram below circle the three parts of the triangle that are labeled. (Assume each triangle is a right triangle.)

    b. Write the trigonometry function you would need to solve for x. Then find the missing part.

1.   12   _____

    5   _____

     x°

2.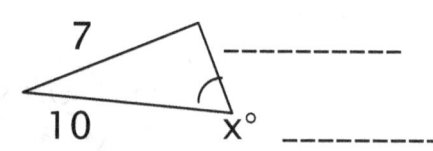

     7   _____

   10   x°   _____

3.   14   x   _____

          35°   _____

4.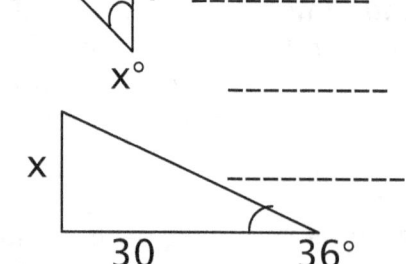

   11   7   _____

      x°   _____

5.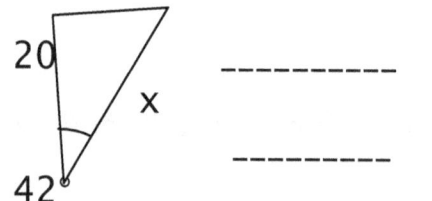

   20   _____

      x

   42°   _____

6.   x   _____

      30   36°

             _____

Tackling Word Problems

Be sure to sketch a diagram that represents the problem first so that you are using the correct function.

Example:

A. The tailgate of a truck is 2 feet above the ground. The incline of a ramp used for loading Kyle's snowmobile on the truck is 11°. Find to the nearest tenth of a foot, the length of the ramp.

B. Adam is snowboarding down a steep hill. Jordan is watching from the bottom of the hill and measures the angle of inclination as 65°. If it takes Adam 2 minutes to descend the hill at an average speed of 35 mph, how high vertically up the hill did Adam start? (In other words, what is the relative altitude of his starting point?)

C. Josh is four-wheeling with friends and is embarrassed when he cannot climb a particular hill. He notices that he always seems to get stuck when hills have a certain steepness. In order to avoid future embarrassment he returns later to measure the hill. From the base of the hill to the point he got stuck is 50 feet. Walking parallel to the hill, he estimates that the horizontal distance is 25 feet. What angle of inclination should he avoid with his four-wheeler? (At least until he saves up to buy one with a more powerful motor!)

D. Rachel is babysitting two children who have a cranky neighbor. They go out to fly kites, but Rachel must be sure that the kites do not fly over the neighbor's property. Under current wind conditions, the kites are flying at a 50° angle and the ideal location is 100 feet from the neighbor's property. At what length should Rachel limit the kites' string?

E. Alex is sitting in his tree stand when he sights an eight-point buck. He wants to wait until the deer is 100 yards away from the tree. He knows his tree stand is 12 feet up in the tree. What angle of depression in his line of vision should he be waiting to fire at?

F. You are trying to find how wide Black River is. You are standing on one side, your friend is on the other. You fly a kite so that it crosses over to the other side. When your friend says that the kite is directly overhead, you measure the angle of elevation as 65° and the kite string has been let out to 220 feet. How wide is the river at this point?

% Error

A measure of how accurate measurements are is the % error.

In science laboratory experiments you may be asked to calculate your % error based on what the instructor knows the value should be.

Read the article "Survey Finds Rockies Higher Than Previously Thought" from the *Watertown Daily Times*. Then complete the table below.

Mountain	Measurements		Difference	Diff./New	x 100	% Error
	Previous	New				
Pikes Peak					x 100	
Mount Elbert					x 100	
Mount Sneffels					x 100	
Mount Evans					x 100	
(Denver, CO)					x 100	

Is this a significant amount of error in measurement?

_____

_____

How much confidence do you have in the new measurements?

_____

"Survey Finds Rockies Higher Than Previously Thought"

Watertown Daily Times
July 23,2002

The Colorado Springs Gazette:

DENVER – People interested in tackling any of Colorado's 14,000-foot mountains better pack an extra energy bar – the hills are taller than anyone thought.

Pikes Peak, for example, is really 5 feet higher than its old measurement of 14,110 feet.

This tall tale comes from the National Geodetic Survey, a federal group that deals with latitudes, longitudes, heights, orientation and shoreline measurements from sea to shining sea.

The National Geodetic Survey recently recalculated measurements across the Rocky Mountains. Mount Elbert, the state's tallest peak, now officially is 14,440 feet, 7 feet higher than previously thought.

Mount Sneffels in southwestern Colorado measures 14,153 feet, an increase of 3 feet from the old mark. Mount Evans also underwent a growth spurt and now stands at 14,265, and increase of a foot.

So are the mighty wonders bulking up on steroids? Hardly.

In fact, most mountains grow at a fraction of a millimeter each year. That's barely enough to compensate for erosion, experts say.

The truth is the old elevation levels – in place since the 1920's – are slightly off. That's been revealed by new methods of calibration that rely on Global Positioning System satellites, not survey equipment.

The bottom line is many old readings, from the state's 54 14,000-foot mountains to a mile-high marker at the state Capitol, are a tad off.

The new measurements won't mean much to the average person, and they won't change the rankings for any of the state's 14,000-foot peaks. Nor will they add any new members to the 14,000-foot club.

But the recalculation shows what progress brings. The old method of calculation, completed in 1929 by the National Geodetic Survey, relied on standard survey tools to determine elevation points.

They were accurate, but not exact.

Global Positioning System technology and new scientific methods used to better measure the Earth's surface have raised the bar, at least in terms of map points.

"All this only matters if you are doing very precise surveying," said Jack Reed, an emeritus scientist with the U.S. Geodetic Survey, who first wrote about the differences in Trail and Timberline, a magazine published by the Colorado Mountain Club.

"If you want to measure the slope of the Platte River, you may need to know it exactly. But for most people it does not make a **** bit of difference."

Besides, the old calculations are not wrong, Reed said. They were based on data that has been usurped by technology.

"It's like in football when you move the down markers," Reed said. "First down depends on where you put the back end of the chain."

So what does this mean for the Mile High City, which now is slightly higher than its famous moniker?

Not much.

"I guess the people in Denver are 3 feet closer to heaven than everyone else," said Andrew Hudson, spokesman for Denver Mayor Wellington Webb.

To heck with the survey. Denver isn't changing its nickname.

"Mile High City is better than Mile-And-Three-Feet City," Hudson said.

The new elevation levels make two mile-high markers in Denver somewhat out of step with the new reality.

A row of purple seats at the upper decks of Coors Field meant to designate the mile-high mark are also 3 feet off.

No one cares.

But don't look for any alteration.

"There are no plans to make a change," said Dave Moore, director of Coors Field administration and development.

Measuring With Trigonometry

In this lesson we will use trigonometric ratios to estimate heights and/or depths.

1. Begin by measuring your average pace. Using a length of step that is even and seems natural to you, walk across the pre-measured area. Record how many steps you take: _____

2. The measured area is _____ feet. How long was your average step (pace)? _____

3. The measuring device you will be using is a crude variation of the astrolabe, the instrument invented by early astronomers to measure the stars. You will need to work in groups of two or three so that one person can view the top of the object through the tube and the other can record the degrees marked by the weighted string. If a third person is in your group they should record the measurements.
    a. For best results, the top of the object should be in the upper third of the viewing tube.
    b. Be sure the string hangs and moves freely.
    c. Use even steps the same length you used in the measured area to "measure" the distance from the object to the point it is being measured from.
    d. Try to stand on ground even with the object being measured. If this is not possible, find a way to account for this difference in height.

4. While outside, only record the object being measured, the number of degrees measured, the number of paces, who paced that distance, and who was holding the measuring instrument. The later two can be recorded by that person initialing the chart.

5. When back inside, complete the chart. Use the complement of the angle. The height adjustment is necessary because

   _____
   _____.

6. Use the _____ function to determine the estimated height.

7. When inside do a "control" measurement. To do this measure the height of the room from the same mark you paced from.

   _____

8. Find the percent error in your control. The actual height of the
   room is _____ feet.

   a. Find the difference between the actual height and the height
      you measured.
   b. Divide this difference by the actual height.
   c. Multiply by 100.
   d. This number is your percent error.

9. Use this percent error to find the likely range of measurements for
   each of your estimates.

10.   Discuss at least three possible causes for error in the
   measurements.

(1) _____

    _____

    _____

    _____

(2) _____

    _____

    _____

(3) _____

    _____

    _____

(4) _____

    _____

    _____

Measuring With Trigonometry

Object:	Angle Measured: Complement:	Paces to Object:	Est. Feet to Object:	Height Adjustment:	Estimated Height:	% Error Range:

Right Triangle Trigonometry

<u>Aug '01, #33:</u> A ship on the ocean surface detects a sunken ship on the ocean floor at an angle of depression of 50°. The distance between the ship on the surface and the sunken ship on the ocean floor is 200 meters. If the ocean floor is level in this area, how far above the ocean floor, to the nearest meter, is the ship on the surface?

<u>Jan '02, #35:</u> Draw and label a diagram of the path of an airplane climbing at an angle of 11° with the ground. Find, to the nearest foot, the ground distance the airplane has traveled when it has attained an altitude of 400 feet.

<u>Jan '01, #35:</u> Find, to the nearest tenth of a foot, the height of the tree represented in the accompanying diagram.

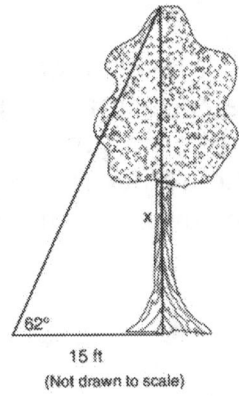

62°

15 ft

(Not drawn to scale)

<u>Aug '99, #27:</u> A person standing on level ground is 2,000 feet away from the foot of a 420-foot-tall building, as shown in the accompanying diagram. To the nearest degree, what is the value of x?

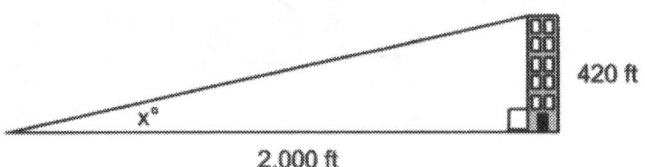

420 ft

x°

2,000 ft

<u>Sample '98, #25:</u> The tailgate of a truck is 2 feet above the ground. The incline of a ramp used for loading the truck is 11°, as shown below.

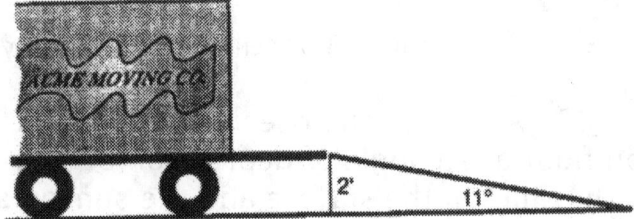

Find, to the nearest tenth of a foot, the length of the ramp.

<u>Aug '00, #33:</u> A 10-foot ladder is to be place against the side of a building. The base of the ladder must be placed at an angle of 72° with the level ground for a secure footing. Find, to the nearest inch, how far the base of the ladder should be from the side of the building and how far up the side of the building the ladder will reach.

<u>June '00, #30:</u> A surveyor needs to determine the distance across the pond shown in the accompanying diagram. She determines that the distance from her position to point P on the south shore of the pond is 175 meters and the angle from her position to point X on the north shore is 32°. Determine the distance, PX, across the pond, rounded to the nearest meter.

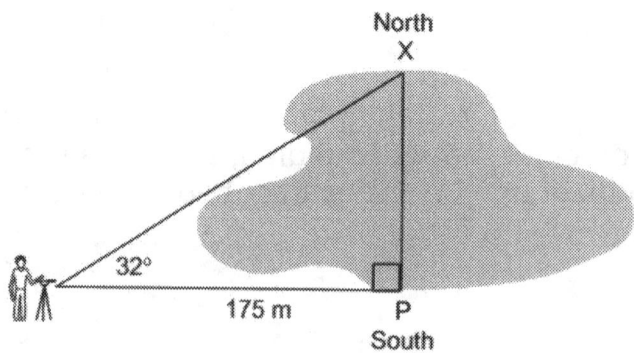

Aug '02, #31: In the accompanying diagram, x represents the length of a ladder that is leaning against a wall of a building, and y represents the distance from the foot of the ladder to the base of the wall. The ladder makes a 60° angle with the ground and reaches a point on the wall 17 feet above the ground. Find the number of feet in x and y.

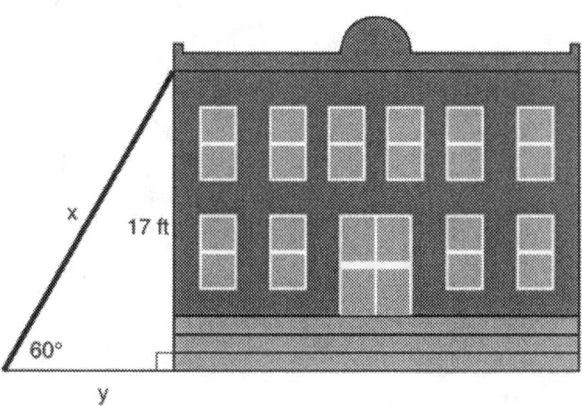

June '02, #31: As seen in the accompanying diagram, a person can travel from New York City to Buffalo by going north 170 miles to Albany and then west 280 miles to Buffalo.

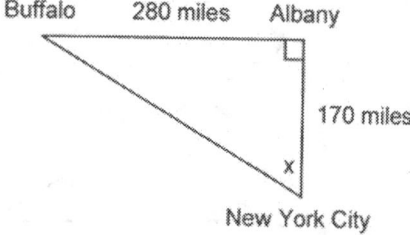

a. If an engineer wants to design a highway to connect New York City directly to Buffalo, at what angle, x, would she need to build the highway? Find the angle to the nearest degree.

b. To the nearest mile, how many miles would be saved by traveling directly from New York City to Buffalo rather than by traveling first to Albany and then to Buffalo?

<u>June '99, #34:</u> Joe is holding his kite string 3 feet above the ground, as shown in the accompanying diagram. The distance between his hand and a point directly under the kite is 95 feet. If the angle of elevation to the kite is 50°, find the height, h, of his kite, to the nearest foot.

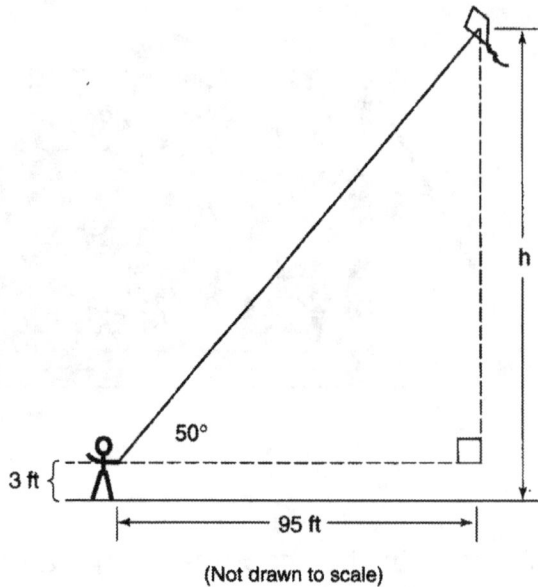

50°

3 ft

|← 95 ft →|

(Not drawn to scale)

<u>Jan '03, #16:</u> In the accompanying diagram of right triangle ABC, AB=8, BC=15, AC=17, and m∠ABC=90.

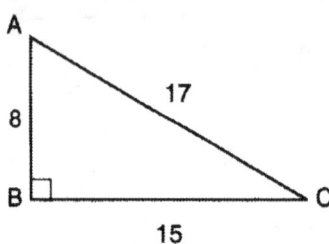

What is tan∠C?

(1) $\dfrac{8}{15}$ (2) $\dfrac{17}{15}$ (3) $\dfrac{8}{17}$ (4) $\dfrac{15}{17}$

# Latitude & Longitude

Latitude and longitude are measurements of the earth in _____

_____ & _____

The earth has an equatorial diameter of 7,926 miles. Therefore, the circumference of the earth at the equator is _____

The earth has a polar diameter of 7,900 miles. Therefore, the circumference of the earth at the poles is _____

Longitude measures how far _____ or _____ a location is from the _____ _____ .

Latitude measures how far _____ or _____ a location is from the _____ .

Latitude is a uniform distance. Each degree of latitude represents 68.708 miles.

The distance each degree of longitude represents varies depending on the latitude. (The longitude lines meet at the poles and are farthest apart at the equator.)

We can approximate the distance a degree represents by
1. Finding the cosine of the latitude
2. Multiplying this number by 69.171 (the miles one degree of longitude represents at the equator).

The table on page 196 contains the latitude and longitude of 9 US cities accurate to the nearest second. On an object as large as the earth we want to be able to pinpoint locations. If a degree represents almost 70 miles in some locations, this would not give precise locations. Therefore, just as a clock can be read more precisely in minutes and seconds, we will look at angle measurements in degrees, minutes, and seconds.

To enter an angle measurement in degrees, minutes, and seconds:
1. Begin by checking to make sure your mode is in degrees.

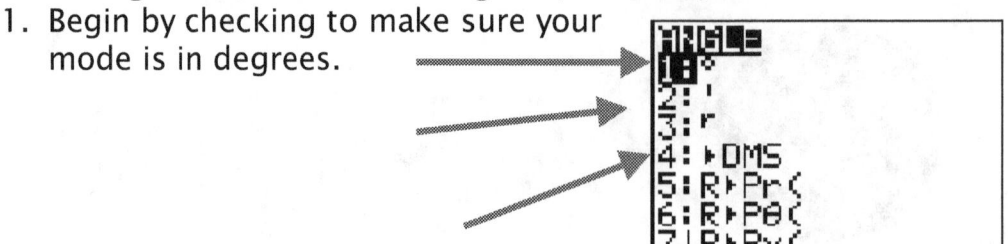

a. Enter the degrees.
b. Enter the degree symbol found by press | 2nd | APPS | and choosing 1:°
c. Enter the minutes.
d. Enter the minute symbol found by press | 2nd | APPS | and choosing 2:'
e. Enter the seconds.
f. Enter the second symbol found by press | ALPHA | + |

For each city in the table, find the approximate distance from your school to that city.

A. To find the vertical distance:
1. Subtract the two latitude coordinates in degree, minute, second form.

2. Convert the decimal answer to degrees, minutes, seconds and enter in the vertical DMS column in the chart.

3. Multiply the DMS by 68.708 (the number of miles in each degree latitude).

4. Round to the nearest mile and enter this value in the Vertical Miles column in the chart.

B. To find the approximate horizontal distance:
1. Subtract the two longitude coordinates and enter this value in Horizontal DMS in the chart.

2. Average the two latitude coordinates.

3. Find the cosine of the average found in (2).

4. Multiply the difference found in (1), the cosine of the average latitude found in (3) and 69.171 (the number of miles per degree longitude at the equator).

5. Round to the nearest mile and enter this value in Horizontal Miles in the chart.

C. To find the approximate straight distance between the two locations, use the horizontal and vertical distances as the legs of a right triangle and use the Pythagorean theorem to find the hypotenuse. Round to the nearest mile and enter in the Distance column in the chart.

How Far Is It?

City	Latitude DMS	Longitude DMS	Horizontal DMS	Horizontal Miles	Vertical DMS	Vertical Miles	Distance
Buffalo, NY	42°31'33"N	78°52'43"W					
Green Bay, WI	44°31'9"N	88°1'11"W					
Miami, FL	25°46'26"N	80°11'38"W					
Los Angeles, CA	34°3'8"N	118°14'34"W					
Nome, AK	64°30'4"N	165°24'23"W					
Syracuse, NY	43°2'53"N	76°8'52W					
Washington, DC	38°53'42"N	77°2'12"W					
Boston, MA	42°21'30"N	71°3'37"W					
Denver, CO	39° 44'21"N	104°59'3"W					
Your School							

Scientific Notation

Scientific notation will be familiar to most of you so this lesson will focus only on how to enter numbers in scientific notation, convert from standard to scientific, and vice-versa, and basic operations with numbers in scientific notation.

First change the calculator from normal mode to scientific.

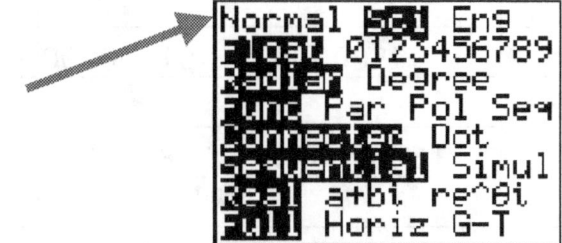

Although the calculator can work with scientific notation in normal mode, in scientific mode every answer will be automatically expressed in scientific notation.

Complete the table on the next page. No calculations should be necessary. With the calculator in scientific mode, input the number, press ENTER, and the calculator should convert the number to scientific notation. Write the number in scientific mode in the box below its normal notation.

Note that the calculator uses the notation 1.34E4 to write 1.34×10^4.

On page 199 you will need to input numbers in this form. To insert the "E" press ⎯2nd⎯ ⎯,⎯ . Although it says EE above the comma key, only one E will appear in scientific mode. The double E is used in engineering mode, which we will not use.

All measurements are in miles.

Planet	Distance from Sun		Distance from Earth		Mean
	Maximum	Minimum	Maximum	Minimum	Radius
Mercury	43400000	28600000	138000000	48000000	1516
Venus	67700000	66800000	162000000	24000000	3760
Earth	94500000	91400000	--------------	--------------	3959
Mars	154900000	128400000	249000000	34000000	2106
Jupiter	507400000	460100000	602000000	366000000	43441
Saturn	941100000	840400000	1031000000	743000000	36184
Uranus	1866400000	1703400000	1962000000	1604000000	15759
Neptune	2824500000	2761600000	2913000000	2676000000	15301
Pluto	4538700000	2755800000	4681000000	2668000000	743
Sun	XXXXXXX	XXXXXXX	XXXXXXX	XXXXXXX	432500
Moon	XXXXXXX	XXXXXXX	225744	251966	1080

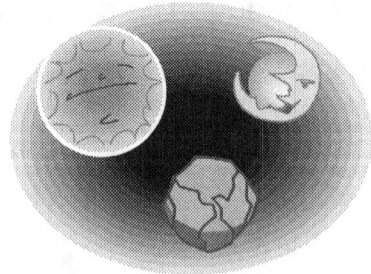

Operations with scientific notation:

Write answers in scientific notation.

1. $1.03 \times 10^5 + 4.07 \times 10^4$

2. $6.75 \times 10^{-5} + 3.69 \times 10^{-6}$

----

3. $4.85 \times 10^{11} + 1.98 \times 10^{10}$

4. $3.08 \times 10^{-34} + 5.97 \times 10^{-30}$

----

5. $5.76 \times 10^{14} - 4.76 \times 10^{15}$

6. $5.82 \times 10^{-12} - 1.34 \times 10^{-13}$

----

7. $6.05 \times 10^3 - 4.05 \times 10^3$

8. $7.15 \times 10^{-1} - 4.78 \times 10^5$

----

9. $8.07 \times 10^{19} \times 1.38 \times 10^{35}$

10. $7.06 \times 10^{43} \times 5.99 \times 10^{-43}$

----

11. $6.55 \times 10^6 \times 4.21 \times 10^7$

12. $5.07 \times 10^{-50} \times 2.31 \times 10^{-7}$

----

13. $7.09 \times 10^{18} \div 4.65 \times 10^9$

14. $3.98 \times 10^{-4} \div 9.76 \times 10^8$

----

15. $\dfrac{2.56 * 10^{11}}{4.21 * 10^{-7}}$

16. $\dfrac{5.55 * 10^{-16}}{4.11 * 10^{-24}}$

Rounding

Remember: **Rounding is not "cutting off".**

The rules for rounding:

1. _____
   _____

2. _____
   _____

3. _____

4. _____

Many regents questions will specify what place to round to.

If the question does not specify, how do you know?

1. Unless there is an implied value to round to DON'T ROUND! Use the entire value seen on your calculator screen.

2. Money is always easy. Round to the nearest cent unless the directions tell you the nearest dollar.

3. As noted above, if the object cannot be cut, you will need to use either the whole number directly above or below. Which one you choose depends on the question, not on the other rounding rules.

4. In science there is a commonly used rule called significant digits or significant figures. (Or for the truly science savvy, "sig figs".)

   An answer cannot be more precise than any of the numbers used to find it!!

   Significant digits are not difficult if you stop to think which numbers are really meaningful.

In the number 1,000,000 the zeroes are not meaningful. They are placeholders. If you want to tell someone you've won exactly one million dollars, it is better to write $1,000,000.00. This implies that the number is accurate to the nearest cent. In general, trailing zeroes with no decimal after them have no real meaning in a number.

Zeroes after a decimal point, but before nonzero digits are also placeholders. In the number .0000025, only the 2 and the 5 have any "significance".

Be careful though!! In the number .00560780, only the first two zeroes are considered "unimportant". If a zero falls at the end of a decimal number it should be included. It tells the reader that the number is accurate to that decimal place.

Tell how many significant digits each of the numbers below has:

a. 250 _____

b. .00315 _____

c. 25060 _____

d. .00983 _____

e. 2.00150 _____

f. 2×10^6 _____

How do sig figs help us to round?

When *adding or subtracting*:
        The answer will have the same number of decimal places as the least accurate number (if one number goes to hundredths and the other goes to thousandths, your answer will be to the nearest hundredth).

When *multiplying or dividing*:
        Count the significant digits in the numbers you use to calculate. Your answer cannot have more significant digits than any of the numbers used to calculate it. (Beware!! The people who write math regents questions don't always follow this rule!! .....When in doubt, don't round!!)

5. Finally, the calculator has a glitch that requires rounding. Occasionally the calculator will give an answer in Solver or on the graphing screen that is impractical. We have seen solutions with repeating 9's and solutions with repeating 0's. These should both be rounded. The last non-repeating digit should be rounded up if the repeating digit is 9, left as is if the repeating digit is 0.

One last item to keep in mind: The following quote is from the NYS Education Department Guidelines for graphing calculator use.

"It is expected that students are capable of performing the following tasks using a graphing calculator:
...Using the full potential of the technology by storing all of the digits produced by the calculator during computation. **Rounding to the specified degree of accuracy should be done only at the end of all computation when the final answer is found.**"

Compute and correctly round each of the following: (Use the sig figs rule.)

1. $3.01 \times 10^5 - 4.9 \times 10^3$ _____ 2. 230,100 x 4.71 _____

3. .009 + 7.8 _____ 4. 6.78/5.27 _____

5. 12,030,001 x .034514 _____ 6. $5.4 \times 10^7 + 3.6 \times 10^3$ _____

7. 5.002 x 4.6 _____ 8. 7,000,000/43.1 _____

9. 89.060 – 34 _____ 10. 2304 – 2290 _____

11. 1.357 x 5.0 _____ 12. $90.8723 \times 10^{-7} - 1.04 \times 10^{-4}$

_____

13. 400,009 x .0212 _____ 14. $1.907 \times 10^3 / 4.68 \times 10^8$

_____

15. 5,678,901,345.007 x 4.007361 _____

Scientific Notation

Jun'00, #29: The distance from Earth to the imaginary planet Med is 1.7×10^7 miles. If a spaceship is capable of traveling 1420 miles per hour, how many days will it take the spaceship to reach the planet Med? Round your answer to the **nearest day**.

Aug '00, #4: Expressed in decimal notation, 4.726×10^{-3} is

(1) 0.004726   (3) 472.6
(2) 0.04726   (4) 4,726

Jan '01, #11: The distance from Earth to the Sun is approximately 93 million miles. A scientist would write that number as

(1) 9.3×10^6 (3) 93×10^7
(2) 9.3×10^7 (4) 93×10^{10}

Aug '99, #4: Which expression is equivalent to 6.02×10^{23}?

(1) 0.602×10^{21} (3) 602×10^{21}
(2) 60.2×10^{21} (4) 6020×10^{21}

Jan '00, #18: If the number of molecules in 1 mole of a substance is 6.02×10^{23}, then the number of molecules in 100 moles is

(1) 6.02×10^{21} (3) 6.02×10^{24}
(2) 6.02×10^{22} (4) 6.02×10^{25}

June '01, #8: If 0.0154 is expressed in the form 1.54×10^n, n is equal to

(1) –2   (3) 3
(2) 2   (4) –3

<u>Jan '02, #6:</u> The approximate number of seconds in a year is 32,000,000. When this number is written in scientific notation, the numerical value of the exponent is

(1) –7   (3) 7

(2) 6   (4) 8

<u>June '02, #7:</u> If 3.85×10^6 is divided by 385×10^4, the result is

(1) 1   (2) 0.01   (3) 3.85×10^2 (4) 3.85×10^{10}

<u>Aug '02, #10:</u> If 0.0347 is written by a scientist in the form 3.47×10^n, the value of n is

(1) –2   (2) 2   (3) 3   (4) –3

<u>Sample #8:</u> If 0.0154 is expressed in the form 1.54×10^n, n is equal to

(1) –2   (2) 2   (3) 3   (4) –3

<u>Jan '03, #19:</u> What is the value of $\dfrac{6.3 \times 10^8}{3 \times 10^4}$ in scientific notation?

(1) 2.1×10^{-2} (2) 2.1×10^2 (3) 2.1×10^{-4} (4) 2.1×10^4

<u>June '03, #1:</u> The number 8.375×10^{-3} is equivalent to

(1) 0.0008375   (2) 0.008375   (3) 0.08375   (4) 8,375

Interpreting Data

The data on the next page was collected using the

_____

This instrument runs on the _____ application for the

_____ .

The _____ is capable of collecting data of the following
types:

_____

_____

_____

_____

_____

_____

The data collected for this lesson used the _____ .

To collect this data, the _____ was programmed to take a

_____ reading every _____ (_____) from

_____ to _____ .

1. Make a scatter plot of the data. Plot both days on the same scatter plot but use different colors or different marks for each day.
2. Mark off the approximate class periods on the x-axis.
3. Find the average temperature for the combined days for each class period and enter this in the chart on page 207.
4. Convert the Celsius mean temperature to a mean Fahrenheit temperature in the next column.
5. Below the chart write a paragraph describing the room temperature (you may especially want to relate it to normal room temperature.)
6. Write a second paragraph discussing why the temperatures might vary throughout the day in the way they did.

12/9/02

L1	L2	L3	1
0	27.163	------	
600	14.125		
1200	14.125		
1800	14.225		
2400	14.525		
3000	14.725		
3600	15.024		

L1(1)=0

L1	L2	L3	1
4200	15.415		
4800	15.707		
5400	16.381		
6000	16.762		
6600	16.667		
7200	16.762		
7800	16.857		

L1(14)=7800

L1	L2	L3	1
8400	17.048		
9000	17.143		
9600	17.429		
10200	17.333		
10800	17.524		
11400	17.619		
12000	17.81		

L1(21)=12000

L1	L2	L3	1
12600	17.429		
13200	17.048		
13800	16.762		
14400	16.857		
15000	16.857		
15600	16.952		
16200	16.952		

L1(28)=16200

L1	L2	L3	1
16800	16.762		
17400	16.762		
18000	17.238		
18600	17.333		
19200	17.619		
19800	17.81		
20400	18.095		

L1(35)=20400

L1	L2	L3	1
21000	18.476		
21600	18.667		
22200	18.667		
22800	18.571		
23400	18.667		
24000	18.667		
24600	18.667		

L1(42)=24600

L1	L2	L3	1
24600	18.667		
25200	18.571		
25800	18.191		
26400	18		
27000	17.81		

L1(47)=

12/10/02

L1	L2	L3	1
0	14.225	0	
600	14.525	------	
1200	14.525		
1800	14.825		
2400	15.317		
3000	15.805		
3600	16.296		

L1(1)=0

L1	L2	L3	1
4200	16.286		
4800	16.381		
5400	16.857		
6000	16.857		
6600	16.952		
7200	17.143		
7800	17.429		

L1(14)=7800

L1	L2	L3	1
8400	17.429		
9000	17.333		
9600	17.524		
10200	17.81		
10800	18.095		
11400	17.905		
12000	17.81		

L1(21)=12000

L1	L2	L3	1
12600	17.429		
13200	17.524		
13800	17.524		
14400	17.333		
15000	17.429		
15600	17.333		
16200	17.429		

L1(28)=16200

L1	L2	L3	1
16800	17.143		
17400	17.429		
18000	17.905		
18600	18		
19200	18.191		
19800	18.667		
20400	18.857		

L1(35)=20400

L1	L2	L3	1
21000	19.048		
21600	19.143		
22200	18.952		
22800	18.952		
23400	19.048		
24000	19.048		
24600	19.048		

L1(42)=24600

L1	L2	L3	1
25200	18.857		
25800	18.667		
26400	18.857		
27000	18.476		

L1(47)=

Time (Class Period)	Mean Celsius Temp.	Mean Fahrenheit Temp.
1		
2		
3		
4		
5		
6		
7		
8		
9		
10		

(5)

----

(6)

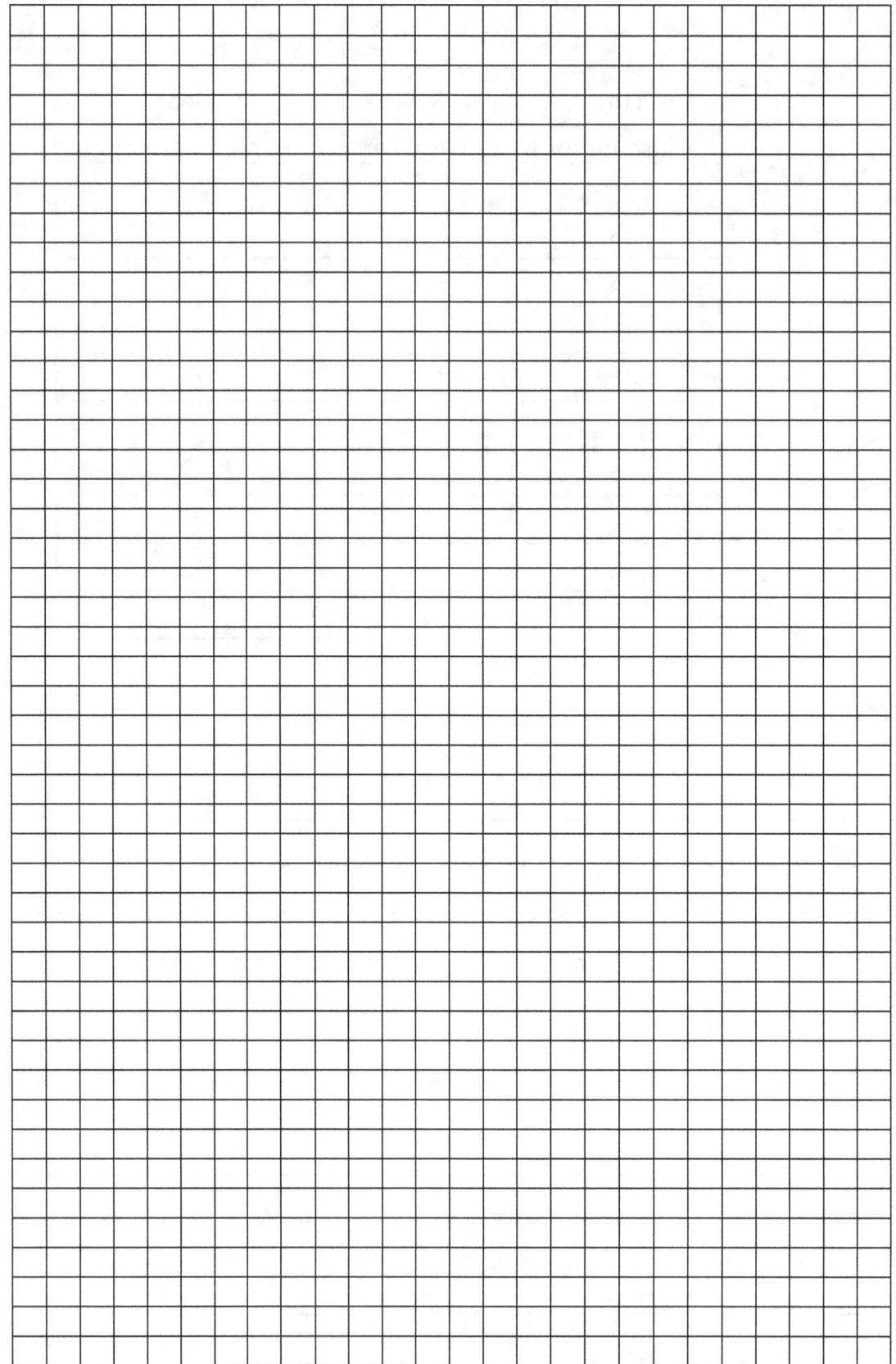

 # Probability

Vocabulary:

Probability _____

_____

_____

A probability experiment is _____

_____

A sample space is _____

_____·

A _____or _____ is when the event in question actually occurs.

You should be especially familiar with the following types of experiments:

1. Rolling dice

2. Drawing cards from a standard deck

3. Picking marbles from a jar or urn

4. Coin flips

5. Spinners

Always assume that a die is a standard six-sided die with each of the six sides equally likely to come up on a roll.

A standard deck of cards has

*52 cards

*4 suits (hearts, diamonds, spades, and clubs)

*13 of each suit (one ace, one each of the

numbers   2 through 10, one Jack, one

Queen, and one King)

*2 colors

*26 of each color

Each of the cards in the deck is equally likely to be drawn.

Picking marbles from a jar will always require a specific set up you must identify from the question.

Assume that every coin is a fair coin and there is an equally likely chance of having the coin land heads or tails.

Spinners are usually shown in the question and will vary depending on the question.

There are two types of probability:

   A. _____

   B. _____

Theoretical probability_____

_____.

Experimental probability _____

_____

Will theoretical usually match the experimental in events in which they can be compared?

_____
_____

Using the TI-83+/TI-84+ we will see how close simulated experiments come to modeling the theoretical probabilities of the common experiments listed above.

Most new calculators come with a probability simulator application. This will be erased if you reset ALL. Linking with a calculator that has the application can easily retrieve it.

See the separate instructions for linking and load the probability simulator on your calculator if it is not already there.

To use the application press [APPS] and choose Prob Sim. It will

have a different number in front depending on how many other applications are currently loaded on the calculator.

Follow the directions on the screen.

You should see a screen like the one at the right.

Notice the 5 boxes at the bottom of the screen. You need to use the top row of keys on the calculator to access these options. In most cases the other keys on your calculator will not function in this application.

Use the key that is in the corresponding position. (Ex. Use the GRAPH key to QUIT.)

These are experiments so it is expected that you will have different answers than your neighbor.

1. Under ideal conditions, how many heads would you expect if you flip a coin 10 times? _____ 50 times? _____ 100 times? _____
   "Flip" the coin 10 times. What is the result? _____

   Note that you can "speed flip" by choosing +10 or +50 after the first toss.

   To see the results of your experiment:
   a. press ESC
   b. choose TABL
   c. CumH gives the cumulative number of heads after each toss.
   d. Use this table to look up the number of heads after 50 tosses: _____

   100 tosses: _____

In the remaining experiments, to view your results at any time during the experiment, press the left or right arrow keys. The frequency should appear above the bar the cursor is on in the bar chart.

2. Go back to the main simulation menu and choose 2. Roll Dice. Under ideal conditions, after 300 rolls you would expect to have:

_____ 1's
_____ 2's
_____ 3's
_____ 4's
_____ 5's
_____ 6's

"Roll" 300 times. Record your results. "Roll" 1200 times. Record your results.

	300:	1200:
1's	_____	_____
2's	_____	_____
3's	_____	_____
4's	_____	_____
5's	_____	_____
6's	_____	_____

3. There are usually an unequal number of types of marbles in an experiment.
Suppose your jar contains 3 red marbles, 5 blue marbles, and 4 green marbles. To simulate this set-up we will call red A, blue B, and green C.

a. In the main simulation menu choose 3. Pick Marbles.
b. Choose SET.
c. Move cursor down to Types:
d. We have three colors so choose 3
e. Go to ADV
f. Under # of Marbles change Marble A to 3, Marble B to 5 and Marble C to 4.
g. Complete the table below:

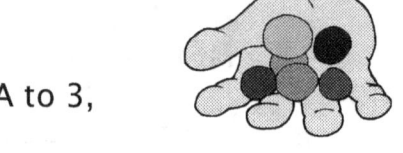

	Expected Outcome:			Actual Outcome:		
	red	blue	green	red	blue	green
10 picks						
100 picks						
1000 picks						

4. Go back to the main simulation menu and choose 4.Spin Spinner.
   a. Choose SET
   b. Change sections to 8
   c. Spin 10 times and record your results
   d. Spin 100 times and record your results
   e. Spin 1000 times and record your results.

	1	2	3	4	5	6	7	8
10 spins								
100 spins								
1000 spins								

5. What is the probability of drawing a black king in 10 draws?_____

   Draw 10 cards. Did you get a black king? _____

   In the default setting the cards are replaced. Is the probability

   better for getting a black king in your next 10 draws? _____

   Explain.

The experimental probabilities should get closer to the theoretical as

the number of trials increases. Did yours? _____ Do they have

to? Explain.

This lesson is intended to be used with the probability simulation activity.

Every TI-83+/TI-84+ comes with a cable that enables you to "link" to another graphing calculator.

Linking is a way to share programs, applications, and lists.

In this lesson we will want to share an application which has been previously erased from the calculator.

To begin the activity, one calculator has retained the probability simulation application that comes with a new TI-83+/TI-84+.
You will first need to receive the application from a calculator that already has it, then send the application on to the next person. This application is also available at the TI website as a free download.

Link the calculators with the cable.
The **receiving calculator** must set up first!!

Press | 2nd | | X,T,Θ,n | Move the cursor to
RECEIVE.

```
SEND RECEIVE
1:Receive
```

The screen to the right should appear.
You should now wait for the sender.

The **sender** should now press

| 2nd | | X,T,Θ,n | . Choose C:Apps...

```
SEND RECEIVE
B:String...
C:Apps...
D:AppVars...
E:Group...
F:SendId
G:SendOS
H:Back Up...
```

Select the application you wish to send. In the example at right *Prob Sim APP is chosen. To SELECT, press ENTER. The triangle shape should change to an arrow.

Highlight TRANSMIT and ENTER.

**If you see this error message check to make sure that the cable is securely connected to both calculators.

Simple programs and lists usually transmit within seconds. More complicated programs and applications can take several minutes. The calculator will tell you when it is done. Do not disconnect the calculators until you see the Done message!

It will also appear if the sending calculator "sends" before the receiving calculator is prepared to "receive". Be sure you are completing the transaction in the proper order.

The Counting Principle

When calculating probabilities, we must first find how many possible outcomes there are.

Sometimes this will be very obvious. Other times we will need to do some calculating.

When we are looking at possible numbers of combinations of different types of objects we can use <u>the counting principle</u>.

<u>The counting principle</u> says that the number of possibilities is the <u>product</u> of the number of objects in each set.

Example: A school's lunch menu allows students to choose one item from each of the following lists:

<u>Beverage:</u>	<u>Sandwich:</u>	<u>Soup:</u>	<u>Dessert:</u>
Milk	Peanut Butter	Tomato	Fruit
Chocolate milk	Tuna	Vegetable	Cookie
Juice	Turkey		Ice
Cream			

Because there are 3 beverages, 3 sandwiches, 2 soups, and 3 desserts, there are 3x3x2x3=54 possible menu combinations.
***This assumes that a student must take one choice from each group.**

Suppose a student may choose not to have a selection from each list. What if the school policy says that they may not choose only a beverage, only a dessert, or only a beverage and dessert.

Now the possible combinations are:

--------------- ------------- -------------- ---------

--------------- ------------- --------------

--------------- ------------- --------------

--------------- ------------- --------------

--------------- ------------- --------------

--------------- --------------

--------------- --------------

--------------- --------------

--------------- --------------

--------------- --------------

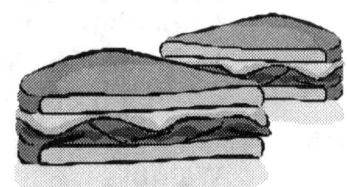

Use the counting principle on each combination and find the sum of all of these combinations.

How many students could eat lunch and no two of them have the same combination of foods? _____

Practice:

1. If Sally has 4 shirts, 2 skirts, and 3 sweaters that are color coordinated, how many different outfits can she put together assuming that she chooses one item from each group?

   _____

2. Mary is wallpapering her bedroom. She has found 5 patterns that she likes. Each pattern has 4 coordinating borders and 9 choices for coordinating paint for the trim. How many combinations does she have to choose from?

   _____

3. In Jim's senior year he can choose from 3 English courses, 2 social studies electives, 3 math courses, 2 science courses, and he has room in his schedule for an elective of his choice from a list of 10. How many possible combinations of courses could he take? (Assume he must take one English, one math, one social studies, and one science course, and he will take another course as an elective.)

   _____

4. An ice cream shop claims to be able to serve 1000 different sundaes.   They offer the following choices:

Ice Cream Flavors:	Toppings:	Extras (choose 2):
Vanilla	Hot Fudge	Chocolate Sprinkles
Chocolate	Chocolate	Rainbow Sprinkles
Strawberry	Caramel	Whipped Cream
Orange Sherbet	Marshmallow	Peanuts
Cookie Dough	Strawberry	Walnuts
Mint Chocolate Chip		Maraschino Cherry

   Are they being honest? Explain.

_____
_____
_____

Compound Events

There are several cases that must be considered when we go beyond simple probability.

When more than one "event" occurs, sometimes the outcome of that event will effect the probability of the "event" paired with it and sometimes it won't.

Independent Events

When the two (or more) events will have no effect on each other they are considered independent.

Ex: Rolling a die, then flipping a coin.

One way to calculate the probability would be to make a tree diagram:

You could also make a list of all the possible outcomes:

We can also calculate the number of outcomes: (Use the counting principle)

Dependent Events

If the events are not exclusive or in any way influence the outcome of other events, they are called Dependent.

Example 1: At the awards assembly 10 students are eligible for 3 awards.
What is the probability that a particular one of the 10 will receive an award?

What information do you need before you can calculate the probability?_____

_____

If one student can receive more than one award, the probability is different than if a student can only receive one award. What is the number of possibilities in each case?
_____

And vs. Or

Which would you expect to be higher, the probability of rolling a 3 *and* a coin landing on heads, or the probability of rolling a 3 *or* a coin landing on heads? _____

Remember that probabilities are _____

Which produces a smaller number, adding two numbers ≤ 1 or multiplying two numbers ≤ 1? _____

If you are looking at the probability that both events have favorable outcomes, multiply the probabilities of the individual events.
If you are looking at the probability that one or the other of two events will have a favorable outcome, add the probabilities but subtract out any overlapping possibilities.

With or Without Replacement

It matters if an object is returned to the set being drawn from if there is a favorable outcome on the first trial.

In the previous example, it mattered if a student was returned to the "pool" of possible award winners.

When replacing the original object, the probabilities are not changed.

Example 2: What is the probability of drawing a black card three times in a row if the card drawn *is replaced* each time?(Black *and* Black *and* Black)

When the first trial results in a favorable outcome and the object is not returned to the set of possible results, there are two possibilities:

      1. Are you looking for another outcome from the same set as the first one,

      2. Or, is a favorable outcome for the second trial different than it was for the first?

In the first case, you will need to subtract one from both the numerator and the denominator of your second probability before multiplying.

In the second case, you will need to subtract only from the denominator.

Example 3: What is the probability of drawing a black queen, then drawing any black card *without replacement?* (Black Queen *and* Black)

Example 4: What is the probability of drawing a black queen, then drawing a red card *without replacement?* (Black Queen *and* Red)

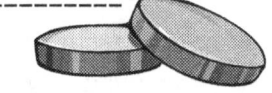

Practice:

      1. What is the probability of a coin landing heads and a die landing on an even number?

      2. What is the probability of a die landing on a prime number and drawing a heart from a standard deck?

      3. What is the probability that a coin will land on tails, or a die will land on a multiple of 3?
         (Remember to subtract out those cases where both are occurring or you will be counting them twice.)

      4. If a package of m&m's contains 15 blue, 20 green, and 13 red m&m's, what is the probability of drawing one of each when picking them out of the package one at a time without looking?

        What is the probability of getting all three m&m's the same color?

Combinations, Permutations, and Factorials

Combinations are:

_____ $_nC_r$

_____

Permutations are:

_____ $_nP_r$

_____

Factorial notation is used for:

_____ !

_____

All of these can be performed on the graphing calculator by pressing

| MATH | then choosing the PRB (probability) menu.

```
MATH NUM CPX PRB
1:rand
2:nPr
3:nCr
4:!
5:randInt(
6:randNorm(
7:randBin(
```

The factorial _notation_ is important to know. When using the factorial option you must input the number first, then go to the probability menu and choose #4.

Practice with these:

a. 3! = _____   b. 5! = _____   c. 7! = _____   d. 9! = _____

How do you know when to use nCr and when to use nPr?

When using the nCr or nPr:
  1. Input "n" (the number of possible choices).
  2. Go to the probability menu and select #2 or 3.
  3. Input "r" (how many out of the possible choices you will be
     selecting).

Practice:

a. $_5C_5$ = _____   b. $_3C_2$ = _____   c. $_7C_3$ = _____   d. $_8C_3$ = _____

e. $_6C_4$ = _____   f. $_{10}C_2$ = _____   g. $_{20}C_{11}$ = _____   h. $_{25}C_{20}$ = _____

i. $_5P_5$ = _____   j. $_3P_2$ = _____   k. $_6P_1$ = _____   l. $_8P_3$ = _____

m. $_6P_4$ = _____   n. $_{10}P_2$ = _____   o. $_{20}P_{11}$ = _____   p. $_{25}P_{20}$ = _____

q. Suppose the bookstore has 10 new books in. They will choose 5 of these to display in the front window. How many combinations of 5 books are possible?

r. A special committee is being set up in the US Senate. 10 senators will be chosen at random in the interest of fairness. How many combinations of senators are possible in this committee?

s. Your grandmother has ordered a dozen tulip bulbs and asks you to plant them around her wishing well. Each bulb is a different color and the wishing well has a square base. The bulbs are not labeled so you randomly plant them evenly around the base of the well with none on the corners. How many color arrangements are possible on any one side of the well?

t. One lottery game asks you to choose 4 numbers 0-99. You win regardless of the order you list your numbers. Another lottery game asks you to list 5 numbers 0-25, but you only win if they are in the exact order the winning numbers are chosen. In which game is your chance of winning better? Explain using the probabilities for each game.

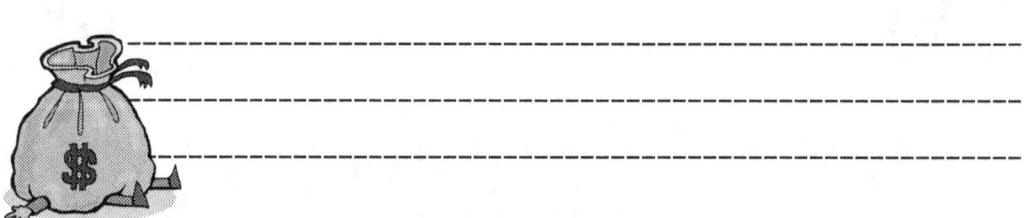

At Most/At Least

When calculating "at most" and "at least" probabilities, we must not only consider the given probability, but also:
   a. all those smaller when it is "at most"

   b. all those larger when it is "at least"

The probability that an event will occur r times out of n trials can be calculated with this formula:

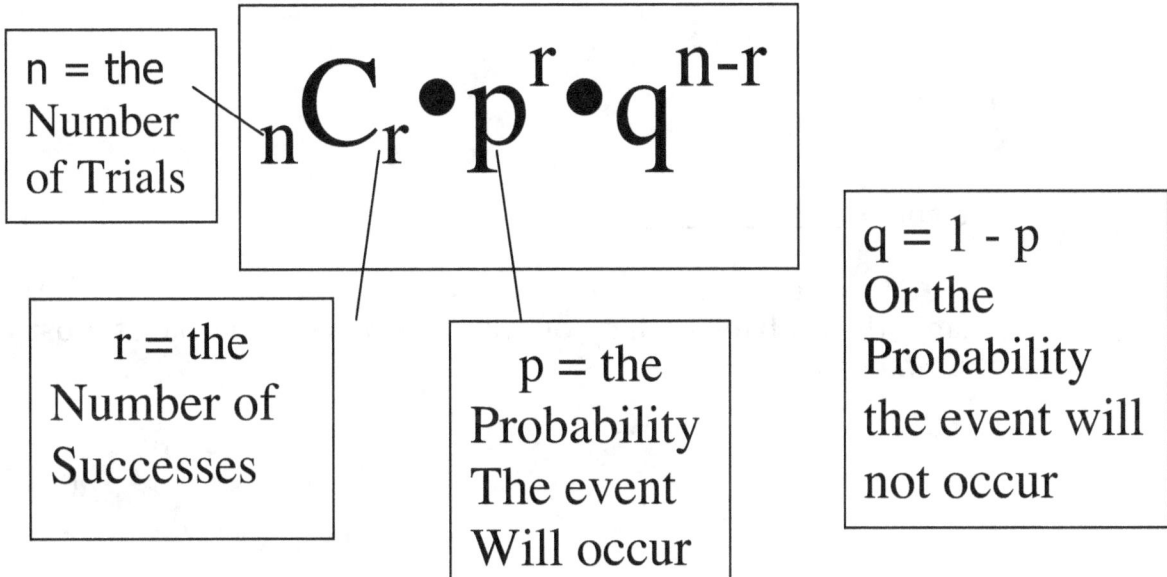

n = the Number of Trials

$$_nC_r \bullet p^r \bullet q^{n-r}$$

r = the Number of Successes

p = the Probability The event Will occur

q = 1 - p
Or the Probability the event will not occur

Let's try just plugging in some numbers first. Suppose n=5, r=3, and p=40%.

$$_nC_r(p^r)(q^{n-r}) = \_\_C\_\_(\_\_\_^{---})(\_\_\_^{---})$$

$$= \_\_\_\_\_ (\_\_\_\_)(\_\_\_\_)$$

$$= _____$$

Suppose a jar contains 3 red marbles, 6 blue marbles, and 2 green marbles. What is the probability of drawing a blue marble at least 3 out of 5 times?

Then you could draw a blue marble ____ times, _____times, or _____ times.

Then you must find $_nC_rp^rq^{n-r}$ for r=3, r=4, and r=5 and add these three probabilities for your final answer.

If r=3:

If r=4:

If r=5:

The sum= _____

What is the probability that you will draw a green marble at most 2 out of 6 times?

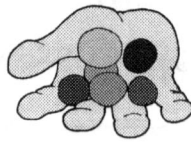

Practice:

1. Suppose your favorite baseball player is batting .365 at the time of a particular game. What is the probability that he will have a base hit at least 3 of 4 at bats in this game?

2. A relief pitcher strikes out 4 out of every 9 batters. The opposing team has the bases loaded with no outs when he enters the game. What is the probability that this pitcher will strike out the next three batters? (At least 3 of 3.)

3. Ann plays varsity basketball and is shooting 60% from the foul line. What is the probability that she will make at least 8 of 10 shots in her next game?

4. A coin has been tested and found to be weighted in such a way that it will land heads 2 of three tosses. What is the probability that it will land heads at most 5 of 10 tosses?

5. What is the probability that a family with 5 children will have at least 2 boys?

6. The weather forecast claims that there is a 50% chance of snow each of the next 5 days. What is the probability that it will snow at most 3 days?

Binomial Theorem

The binomial theorem allows us to use $_nC_r$ to save us work when expanding a binomial raised to a power.

Recall: A binomial is _____

_____

What is $(x+y)^2$?

What is $(x+y)^3$?

As the power on the binomial increases, it becomes more tedious to multiply.

The binomial theorem says:

$$(x+y)^n =\ _nC_0x^ny^0 +\ _nC_1x^{n-1}y^1 +\ _nC_2x^{n-2}y^2 + ... +\ _nC_nx^0y^n$$

Try expanding $(x+y)^3$.

Now expand $(x+y)^4$.

How about $(x+y)^{10}$?

Practice:

1. Expand $(x+y)^7$

2. Expand $(x+1)^3$

3. Expand $(2+y)^5$

4. Expand $(x+7)^2$. Show **_two_** methods.

5. Expand $(x-5)^4$.

6. What is the 4th term of the expansion of $(x+3)^8$?

7. What is the 3rd term of the expansion of $(x+2y)^6$?

8. What is the 2nd term of the expansion of $(7-y)^4$?

Probability

June '99, #1: A fair coin is thrown in the air four times. If the coin lands with the head up on the first three tosses, what is the probability that the coin will land with the head up on the fourth toss?

(1) 0   (2) 1/16   (3) 1/8   (4) ½

June '99, #7: How many different three-member teams can be formed from six students?

(1) 20   (2) 120   (3) 216   (4) 720

June '99, #32: A bookshelf contains six mysteries and three biographies. Two books are selected at random without replacement.
     a.)   What is the probability that both books are mysteries?

     b.)   What is the probability that one book is a mystery and the other is a biography?

Jan '00, #13: How many different 4-letter arrangements can be formed using the letters of the word "JUMP," if each letter is used only once?

(1) 24   (2) 16   (3) 12   (4) 4

Aug '01, #27: There are four students, all of different heights, who are to be randomly arranged in a line. What is the probability that the tallest student will be first in line and the shortest student will be last in line?

Jan '02, #31: A square dartboard is represented on the accompanying diagram. The entire dartboard is the first quadrant from x=0 to 6 and from y=0 to 6. A triangular region on the dartboard is enclosed by the graphs of the equations y=2, x=6, and y=x. Find the probability that a dart that randomly hits the dartboard will land in the triangular region formed by the three lines.

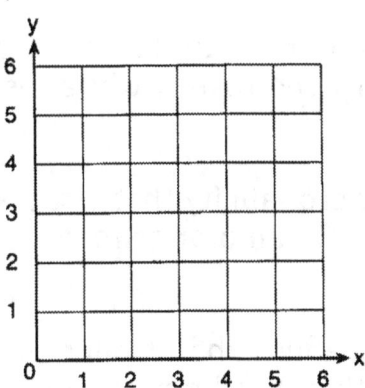

Jan '00, #17: The party registration of the voters in Jonesville is shown in the table below.

Registered Voters in Jonesville	
Party Registration	Number of Voters Registered
Democrat	6,000
Republican	5,300
Independent	3,700

If one of the registered Jonesville voters is selected at random, what is the probability that the person selected is not a Democrat?

(1) 0.333   (2) 0.400   (3) 0.600   (4) 0.667

Sample '98, #20: Erica cannot remember the correct order of the four digits in her ID number. She does remember that the ID number contains the digits 1,2,5, and 9. What is the probability that the first three digits of Erica's ID number will all be odd numbers?

(1) ¼   (2) 1/3   (3) ½   (4) ¾

Jan '00, #34: Three roses will be selected from a flower vase. The florist has 1 red rose, 1 white rose, 1 yellow rose, 1 orange rose, and 1 pink rose from which to choose.

    a.)   How many different 3-rose selections can be formed from the 5 roses?

    b.)   What is the probability that 3 roses selected at random will contain 1 red rose, 1 white rose, and 1 pink rose?

    c.)   What is the probability that 3 roses selected at random will not contain an orange rose?

June '02, #2: If the probability that it will rain on Thursday is 5/6, what is the probability that it will not rain on Thursday?

    (1) 1   (2) 0   (3) 1/6   (4) 5/6

June '02, #34: Alexi's wallet contains four $1 bills, three $5 bills, and one $10 bill. If Alexi randomly removes two bills without replacement, determine whether the probability that the bills will total $15 is greater than the probability that the bills will total $2.

Aug '00, #11: A box contains six black balls and four white balls. What is the probability of selecting a black ball at random from the box?

    (1) 1/10   (2) 6/10   (3) 4/6   (4) 6/4

June '01, #30: Mr. Yee has 10 boys and 15 girls in his mathematics class. If he chooses two students at random to work on the blackboard, what is the probability that both students chosen are girls?

<u>Sample '98, #21:</u> The graph below shows the hair colors of all the students in a class.

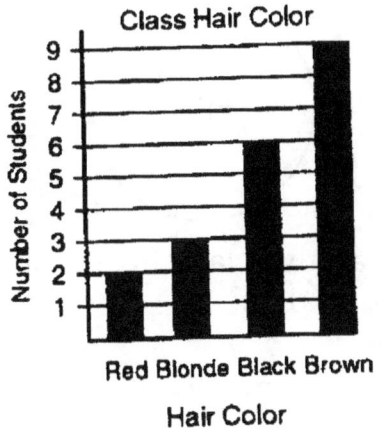

What is the probability that a student chosen at random from this class has black hair?

<u>Jan '02, #9:</u> A fair coin is tossed three times. What is the probability that the coin will land tails up on the second toss?

    (1) 1/3   (2) ½   (3) 2/3   (4) ¾

<u>Jan '01, #26:</u> Sal has a small bag of candy containing three green candies and two red candies.   While waiting for the bus, he ate two candies out of the bag, one after another, without looking. What is the probability that both candies were the same color?

<u>June '00, #34:</u> Paul orders a pizza. Chef Carl randomly chooses two different toppings to put on the pizza from the following: pepperoni, onion, sausage, mushrooms, and anchovies. If Paul will not eat pizza with mushrooms, determine the probability that Paul will not eat the pizza Chef Carl has made.

<u>Jan '01, #6:</u> At a school fair, the spinner represented in the accompanying diagram is spun twice.

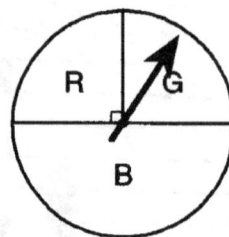

What is the probability that it will land in section G the first time and then in section B the second time?

(1) $\dfrac{1}{2}$ (2) $\dfrac{1}{4}$ (3) $\dfrac{1}{8}$ (4) $\dfrac{1}{16}$

<u>Jan '02, #18:</u> When Kimberly bought her new car, she found that there were 72 different ways her car could be equipped. Her choices included four choices of engine and three choices of transmission. If her only other choice was color, how many choices of color did she have?

(1) 6   (2) 12   (3) 60   (4) 65

<u>Aug '01, #26:</u> Megan decides to go out to eat. The menu at the restaurant has four appetizers, three soups, seven entrees, and five desserts. If Megan decides to order an appetizer or a soup, and one entrée, and two different desserts, how many different choices can she make?

<u>Aug '99, #23:</u> Paloma has 3 jackets, 6 scarves, and 4 hats. Determine the number of different outfits consisting of a jacket, a scarf, and a hat that Paloma can wear.

<u>Aug '01, #11:</u> A certain car comes in three body styles with a choice of two engines, a choice of two transmissions, and a choice of six colors. What is the minimum number of cars a dealer must stock to have one car of every possible combination?

<u>Aug '99, #22:</u> The Grimaldis have three children born in different years.

    a. Draw a tree diagram of list a sample space to show all the possible arrangements of boy and girl children in the Grimaldi family.

    b. Using your information from part a, what is the probability that the Grimaldis have three boys?

<u>June '01, #14:</u> If there are four teams in a league, how many games will have to be played so that each team plays every other team once?

    (1) 6   (2) 8   (3) 3   (4) 16

<u>Aug '00, #25:</u> Alan, Becky, Jesus, and Mariah are four students in the chess club. If two of these students will be selected to represent the school at a national convention, how many combinations of two students are possible?

<u>Aug '01, #7:</u> The value of 5! is

    (1) $\frac{1}{5}$ (2) 5   (3) 20   (4) 120

<u>Aug '99, #17:</u> How many different 6-letter arrangements can be formed using the letters in the word "ABSENT," if each letter is used only once?

    (1) 6   (2) 36   (3) 720   (4) 46,656

<u>Jan '01, #14:</u> A locker combination system uses three digits from 0 to 9. How many different three-digit combinations with no digit repeated are possible?

    (1) 30   (2) 504   (3) 720   (4) 1,000

<u>June '00, #16:</u> How many different five-digit numbers can be formed from the digits 1,2,3,4, and 5 if each digit is used only once?

    (1) 120   (2) 60   (3) 24   (4) 20

June '00, #23: All seven-digit telephone numbers in a town begin with 245. How many telephone numbers may be assigned in the town if the last four digits do not begin or end in a zero?

June '01, #25: There were seven students running in a race. How many different arrangements of first, second, and third place are possible?

Aug '00, #34: The telephone company has run out of seven-digit telephone numbers for an area code. To fix this problem, the telephone company will introduce a new area code. Find the number of new seven-digit telephone numbers that will be generated for the new area code if both of the following conditions must be met:
  • The first digit cannot be a zero or a one.
  • The first three digits cannot be the emergency number (911) or the number used for information (411).

Aug '02, #4: Juan has three blue shirts, two green shirts, seven red shirts, five pairs of denim pants, and two pairs of khaki pants. How many different outfits consisting of one shirt and one pair of pants are possible?

(1) 19   (2) 84   (3) 130   (4) 420

Aug '02, #29: On a bookshelf, there are five different mystery books and six different biographies. How many different sets of four books can Emilio choose if two of the books must be mystery books and two of the books must be biographies?

Jan '03, #7: There are 12 people on a basketball team, and the coach needs to choose 5 to put into a game. How many different possible ways can the coach choose a team of 5 if each person has an equal chance of being selected?

(1) $_{12}P_5$ (2) $_5P_{12}$ (3) $_{12}C_5$ (4) $_5C_{12}$

<u>Jan '03, #21</u>: If Laquisha can enter school by any one of three doors, and the school has two staircases to the second floor, in how many different ways can Laquisha reach a room on the second floor? Justify your answer by drawing a tree diagram or listing a sample space.

<u>Jan '03, #23</u>: Six members of a school's varsity tennis team will march in a parade. How many different ways can the players be lined up if Angela, the team captain, is always at the front of the line?

<u>June '03, #5</u>: Bob and Laquisha have volunteered to serve on the Junior Prom Committee. The names of twenty volunteers, including Bob and Laquisha, are put into a bowl. If two names are randomly drawn from the bowl without replacement, what is the probability that Bob's name will be drawn first and Laquisha's name will be drawn second?

(1) $\dfrac{1}{20} \cdot \dfrac{1}{20}$

(2) $\dfrac{1}{20} \cdot \dfrac{1}{19}$

(3) $\dfrac{2}{20}$

(4) $\dfrac{2}{20!}$

<u>June '03, #20</u>: How many different five-member teams can be made from a group of eight students, if each student has an equal chance of being chosen?

(1) 40   (2) 56   (3) 336   (4) 6,720

June '03, #29: A certain state is considering changing the arrangement of letters and numbers on its license plates. The two options the state is considering are:

Option 1: three letters followed by a four-digit number with repetition of both letters and digits allowed
Option 2: four letters followed by a three-digit number without repetition of either letters or digits

[Zero may be chosen as the first digit of the number in either option.]

Which option will enable the state to issue more license plates? How many more different license plates will that option yield?

June '01B, #22: At a certain intersection, the light for eastbound traffic is red for 15 seconds, yellow for 5 seconds, and green for 30 seconds. Find, to the nearest tenth, the probability that out of the next eight eastbound cars that arrive randomly at the light, exactly three will be stopped by a red light.

Aug '01B, #20: A box contains one 2-inch rod, one 3-inch rod, one 4-inch rod, and one 5-inch rod. What is the maximum number of different triangles that can be made using these rods as sides?

(1) 1   (2) 2   (3) 3   (4) 4

<u>Aug '01B, #28:</u> As shown in the accompanying diagram, a circular target with a radius of 9 inches has a bull's-eye that has a radius of 3 inches. If five arrows randomly hit the target, what is the probability that at least four hit the bull's eye?

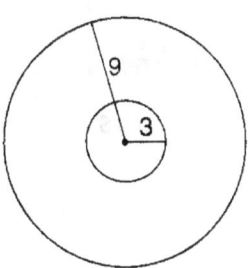

<u>Jan '02B, #29:</u> Team A and team B are playing in a league. They will play each other five times. If the probability that team A wins a game is 1/3, what is the probability that team A will win at least three of the five games?

<u>Samp B, #18:</u> A fair coin is tossed 5 times. What is the probability that it lands tails up exactly 3 times?

   (1) $(1/2)^3$ (2) 3/5   (3) $10(1/2)^5$ (4) $10(1/2)^3$

<u>Samp B, #22:</u> Jim can drive a golf ball over 220 yards 40% of the time. He regularly plays on a golf course where drives of that distance are needed on 12 holes. Determine the probability that exactly 7 of his drives will be over 220 yards.

Aug '02B, #1: Which fraction represents the probability of obtaining exactly eight heads in ten tosses of a fair coin?

(1) $\dfrac{45}{1024}$ (2) $\dfrac{64}{1024}$ (3) $\dfrac{90}{1024}$ (4) $\dfrac{180}{1024}$

Aug '02B, #13: If $_nC_r$ represents the number of combinations of n items taken r at a time, what is the value of $\displaystyle\sum_{r=1}^{3} {}_4C_r$?

(1) 24 (2) 14 (3) 6 (4) 4

Aug '02B, #8: What is the last term in the expansion of $(x+2y)^5$?

(1) y^5 (2) $2y^5$ (3) $10y^5$ (4) $32y^5$

Review 1

1. Reset the calculator. Have the teacher initial this blank before going on to the rest of the review. _____

2. Simplify: 3^5 _____

3. Find -3^2 and $(-3)^2$ _____ _____

Why don't these expressions have the same value?

_____

_____

4. $1\frac{5}{8} + 3\frac{1}{11}$ _____

5. $5\frac{2}{3} - 1\frac{1}{2}$ _____

6. $\frac{2}{8} \cdot 4\frac{1}{7}$ _____

7. Solve for x: $3x - 11 = 42$ _____

8. Solve for x: $9 - 8x = 3$ _____

9. $\frac{1}{3}x + 2 = -10$ _____

10. Find the LCM: 14,15,16 _____

11. Find the LCM: 3,11,20 _____

12. Find the LCM: 5,8,12 _____

13. Find the LCM: 7,19,2 _____

14. Find the GCF: 102,153,357 _____

15. Find the GCF: 121,363,55 _____

16. Find the GCF: 39,44,52 _____

17. Solve: $\frac{30}{x} = 7 + \frac{18}{2x}$ _____

18.   Solve: $\dfrac{4x+6}{x+1} = 5$ _____

19.   Solve: $\dfrac{2}{3x-4} = \dfrac{1}{4}$ _____

20.   Solve: $\dfrac{1}{x} = \dfrac{7}{3x+1}$ _____

21.   Solve: $\dfrac{6}{3x-1} = \dfrac{3}{4}$ _____

22.   Solve: $\dfrac{4x}{7+5x} = \dfrac{1}{3}$ _____

23.   Circle all Pythagorean triples:

      10,24,26   6,8,10

18,24,30   4, $\sqrt{20}$,6

      7,8,12   10,15,20

24. Find the measure of the diagonal of a rectangle whose sides measure 14mm and 48mm.

_____

25. A rectangle has a diagonal of length 10ft and one side of length 6ft. Find the perimeter of the rectangle.

_____

26. A girl 5 ft tall casts a shadow 8 ft long. At the same time, a tree casts a shadow of length 48 ft. What is the height of the tree?

_____

27. The sides of a triangle have lengths 4, 6, 8. In a similar triangle, the shortest side has length x-2 and the longest side has length x+6. Find the value of x.

_____

28. Trina paid $48 for a jacket that was on sale for 30% of the original price. What was the original price of the jacket?

_____

29. 30% of 20 is the same as what % of 24?

_____

30. The profits of a business are to be shared by the three partners in the ratio of 4 to 3 to 2. The profit for the year was $184,662. Determine the number of dollars each partner is to receive.

_____
_____
_____

31. Janet has a ladder that is 17 feet long. If she sets the base of the ladder on level ground 8 feet from the side of the house, how many feet above the ground will the top of the ladder be when it rests against the house?

_____

32. Explain the difference between a rational number and an irrational number.

    ----

33. Name an advantage to resetting the calculator.

    ----

34. If you try to solve $2 + \dfrac{3}{5x} = 2x - 7$ in SOLVER you might get what

    type of error?

    (At least two possible correct answers!)

35. How would you "fix" the error in #34?

    ----

36. Explain a common cause of a "SYNTAX" error.

Review II

Try to answer the following using only the graphing calculator.
You may use your notes.

1. $2\frac{3}{8} - 1\frac{1}{3}$ _____

2. $\frac{4}{7} + \frac{1}{9}$ _____

3. Simplify: $\sqrt{75} + \sqrt{27}$ _____

4. Simplify: $\sqrt{75 + 27}$ _____

5. Solve for x: $4x-3=20$ _____

6. Find the LCM: 3,5,25 _____

7. Find the GCF: 28,35,98 _____

8. If the legs of a right triangle are 4 and 10, what is the hypotenuse?

_____

9. Solve for x: $\frac{x-6}{3x+9} = \frac{3}{5}$ _____

10. Change .456 to a fraction _____

11. Change 3.42 to a mixed number _____

12. What is the value of x in the equation $\frac{3}{5}x + 2 = \frac{7}{4}x - 6$

_____

13. What % of 125 is 10? _____

14. 35 is 74% of what number? _____

15. 43 is what % of 213? _____

16. Sterling silver is made of an alloy of silver to copper in the ratio of 37:3. If the mass of a sterling silver ingot is 720 grams, how much silver does it contain?

_____

17. What is the value of x^3-8x+4 for x=-2? _____

18. What is the value of $|x^5-9|$ for x=-3? _____

19. The grades on a Math A quiz were: 78, 92, 76, 88, 92, 100, 95, 66, 59, 70, 85, 92, 88, 68, 60, 84, 100, 82, 80, 78. Find the following:

Mean: _____

Mode: _____

Minimum X: _____

1st Quartile: _____

Median: _____

3rd Quartile: _____

Maximum X: _____

20. Create a box-and-whisker plot using the data in #19.

21.   Create a frequency histogram for the data in #19.

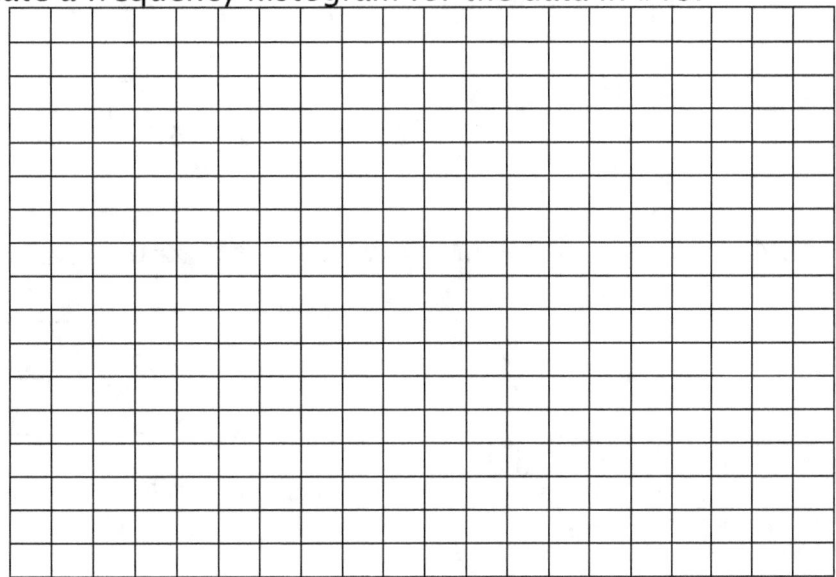

22.   What is the difference between a frequency histogram and a cumulative frequency histogram?

    _____

    _____

23.   What is the difference between a histogram and a bar graph?

    _____

    _____

24.   In the set of positive integers, what is the solution set of the inequality $3x - 8 < 4$?
    (1) {1,2,3,4}   (2) {0,1,2,3}   (3) {1,2,3}   (4) {0,1,2}

21. Expressed in simplest form, $\frac{x}{2} + \frac{x}{3} - \frac{x}{4}$ is equivalent to

    (1) $\frac{x}{9}$ (2) $\frac{x}{24}$ (3) $\frac{13x}{24}$ (4) $\frac{7x}{12}$

(Use test menu!)

22. Find $\left| x^5 + 8 \right| - 3$ when

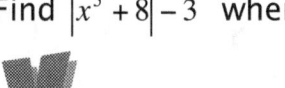

                                              x=-5 _____

                                              x=-4 _____

                                              x=-3 _____

23. Find $\left| x^3 - 2x \right|$ when

      x=-3 _____

      x=-2 _____

      x=-1 _____

24. Find $x^3 - 8x^2 + 3x - 5$ when

      x=0 _____

      x=1 _____

      x=2 _____

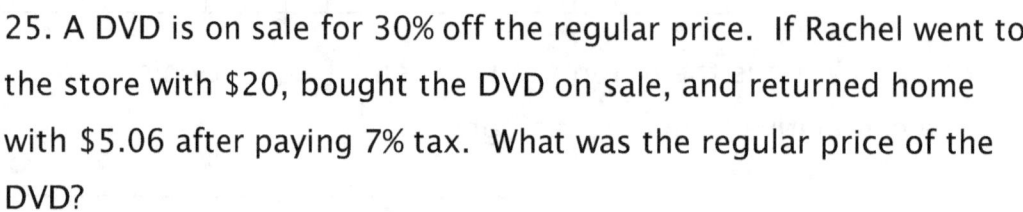

      x=3 _____

      x=4 _____

25. A DVD is on sale for 30% off the regular price. If Rachel went to the store with $20, bought the DVD on sale, and returned home with $5.06 after paying 7% tax. What was the regular price of the DVD?

26. Tom wants to climb to the top of his store roof. He wants to set the ladder at least 6 feet away from the outer wall of the building and the roof is 20 feet from the ground. What length ladder will he need?

Express decimal answers to the nearest hundredth unless otherwise specified.

1. –2^4   _____

2. (-2)^4   _____

3. Add: $\dfrac{4}{5} + 5\dfrac{1}{9}$ (express answer as a mixed number)

           _____

4. Subtract: $18\dfrac{2}{3} - 11\dfrac{7}{9}$ (express answer as a mixed number)

           _____

5. Solve: 4x-21=-42   _____

6. Solve: $\dfrac{8x-10}{x+3} = \dfrac{8}{9}$ _____

7. Express the answer to #6 as a fraction or mixed number.

           _____

8. A 13 foot tree casts a 5 foot shadow. How many feet tall, to the nearest tenth of a foot, is a tree nearby that casts a 7 foot shadow.

           _____

9. What is 14% of 35?   _____

10.   9 is what percent of 45?   _____

11.   If ABC is a right triangle with side a=10cm and side b=14cm, what is the length of the hypotenuse?

           _____

12.   Can the unrounded answer to #11 be converted to a fraction? Why or why not?

     _____

13.   Is the set {150, 200, 250} a Pythagorean triple?

           _____

14. A construction worker has a board that is 10 feet long and is about to prop it against the wall outside Mrs. Noftsier's window. He places the end of the board 7 feet from the outside wall and begins to release the board.   If you estimate that the top of the window is 8 feet up the wall, decide whether or not you should be warning your friend in the seat beside the window to move. (Use the Pythagorean Theorem!). Explain why or why not. (Only mathematical responses please!)

----

15. Find the LCM: 8, 144, 44

----

16. Find the GCF: 1000, 1225, 555

----

17. What is the value of $|x^3-8x|$ for $x=-4$?

----

18. Where on the calculator can you go to check whether a mathematical statement is true or false?

----

19. Test: $(7\cdot12=96) \wedge (7\ne4\cdot2-3)$ ----------

20. What one syllable word does the symbol $\wedge$ in #19 stand for?   ----------

21. The symbol for a disjunction is: ----------

22.   Using the data below, fill in the table.
        Advertising spending for cars, accessories, and equipment
        for the year 1999 (in thousands of $):
        Magazines: $1,835,927.3
        Sunday Magazines: $32,178.2
        Newspapers: $1,544,638.2
        Network Television: $2,573,474.8
        Spot Television: $3,451,710.2
        Syndicated Television: $211,398.5
        Cable Network: $784,641.7
        Radio: $20,826.9
        Source: The World Almanac 2002

MinX	
Q1	
Median	
Q3	
MaxX	

23.   Create a box-and-whisker plot using the data from #22.

24.   What is one inference that can be made from the data in
        #22?

        _____
        _____
        _____

25.   What advantage is there in graphing in zoom decimal or zoom square rather than zoom standard?

_____

_____

_____

26.   Find the equation of the line through (3,5) and (2,-9).

_____

27.   Find the equation of the line through (9,0) and (5,0).

_____

28.   Find the equation of the line through (.5, 1) and (.75,8).

_____

29.   Find the equation of the line through (.5,1) and (.5,10).

_____

30.   Explain why your answers to #27 and 29 are equations of

    a different form:

_____

_____

_____

31. The list at the right gives the troop strength of the countries with the top 20 numbers of active troops.

    a. Enter the number of active troops in L_1.

    b. Sort the list.

    c. Name the 5 countries with the largest numbers of active troops:

    ----

    d. Run 1-Variable Statistics on L_1.

        Mean:_____

        Minx: _____

        Q1:   _____

        Median: _____

        Q3:   _____

        MaxX: _____

Country	Active Troops	Reserve Troops
Afghanistan	400000	NA
China	2820000	1200000
Egypt	450000	254000
Ethiopia	325500	NA
France	317300	416000
Germany	332800	344700
India	1173000	528400
Iran	545600	350000
Iraq	429000	650000
Myanmar	343800	NA
N. Korea	1055000	4700000
Pakistan	587000	513000
Russia	1004100	2400000
S. Korea	672000	4500000
Syria	316000	396000
Taiwan	376000	1657500
Turkey	639000	3787000
Ukraine	311400	1000000
United States	1371500	1303300
Vietnam	484000	3000000

Source: The 2002 World Almanac

e. Create a box-and-whisker plot of the data.

32.   Quiz scores for a class of 15 students were: 88, 60, 75, 86, 92, 100, 67, 85, 89, 75, 90, 70, 82, 78, 93. Make a frequency histogram describing these scores:

Review IV

Unless otherwise specified, round answers to the nearest hundredth.

1. $4\frac{2}{3} - 3\frac{1}{8}$ (Express answer as a fraction.)   _____

2. $5\frac{1}{4} \cdot 2\frac{5}{9}$ (Express answer as a mixed number.)   _____

3. Solve for x: 12x-18=28   _____

4. Express the answer to #3 as a fraction.   _____

5. Solve for x: $\dfrac{2x-8}{5} = \dfrac{3x+4}{-3}$ _____

6. Express the answer to #5 as a fraction.   _____

7. Solve for x: $\dfrac{3x+3}{3} = \dfrac{7x-1}{5}$ _____

8. Solve for x: $\dfrac{4}{x} = \dfrac{6}{x+3}$ _____

9. A woman has a ladder that is 12 feet long. If she sets the base of the ladder on level ground 4 feet from the side of a house, how many feet above the ground will the top of the ladder be when it rests against the house?

    _____

10.   In a right triangle, one leg is 15 cm and the hypotenuse is 32 cm. Find the length of the other leg to the nearest millimeter.

    _____

11.   Find the GCF: 36,54,96   _____

12.   Find the LCM: 7,8,15   _____

13.   What is the value of $\left|3x^5 - 20\right|$ for x=-2?   _____

14.   The scores for a Math A quiz were:
        98, 72, 89, 63, 77, 100, 83, 89, 92, 60, 85, 76, 85, 67,
        85, 95, 83, 75, 70, 90

        Fill in the following table for the quiz scores:

Mean	
MinX	
Q1	
Median	
Q3	
MaxX	

15.   Using the sorted list from #14 make a stem and leaf plot
        for the data.

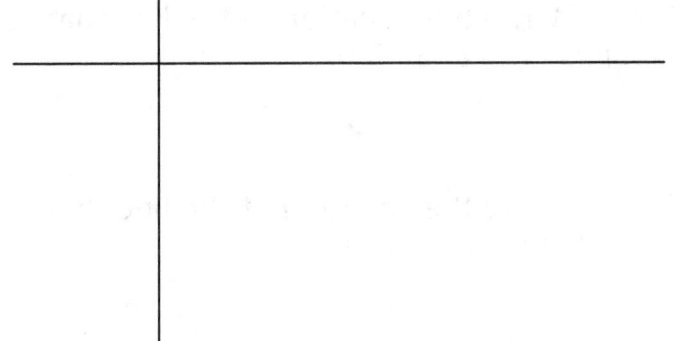

16.   Find the root(s) of the equation $2x^2+4x-20=0$

                                                _____

17.   Find the solution set: $2x^2-10x+8=0$ _____

18.   Find the solution set: $3x^2-5x=0$ _____

19.   Find the solution set: $4x^2-25=0$ _____

20.   Find the solution set: $x^2-8x+16=0$ _____

21.   Find three consecutive integers such that the product of the second and the third, decreased by 4 times the second, is 5 more than 5 times the first.

----

22.   What is the solution set of the equation $\sqrt{x+12} = x$?

----

23. What is the slope of the line passing through the points (3,5) and (7,10)?

----

24. What is the slope of the line passing through the points (4,0) and (4,5)?

----

25.   Write the equation of the line that passes through the points (3,9) and (5,-2).

----

26.   Write the equation of the line that passes through the points (4,2) and (-3,0).

----

27.   Write the equation of the line that passes through the points (2,1) and (2, 10).

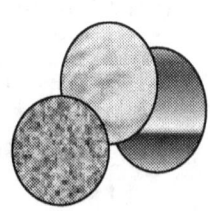

Review V

Unless otherwise specified, round answers to the nearest hundredth.

1. Add: $17\dfrac{13}{27} + 38\dfrac{15}{17}$ (Write answer as a mixed number!) _____

2. Multiply: $1\dfrac{2}{9} * 22\dfrac{11}{13}$ (Write answer as a mixed number!) _____

3. Solve: $\dfrac{8x-1}{3} = \dfrac{15x+2}{7}$ (Write answer as a decimal.) _____

4. Convert the **unrounded** answer to #3 to a fraction. _____

5. Find the LCM: 16, 22, 55 _____

6. Find the GCF: 214, 702, 9005 _____

7. If the legs of a right triangle are 17 and 23, what is the hypotenuse?

    _____

8. What is 14% of 57? _____

9. 42 is what % of 144? _____

10. A DVD you've been waiting for is finally on sale. One store has a regular price of $19.99, but has a 15% storewide sale coming up next week. Another store has a regular price of $18.50 and has a 20% off coupon in this week's newspaper. Which is the better deal and how much would you save by using this deal?

    _____

11. Solve: 117x-14=13x+68 (Write answer as a fraction.)

    _____

12. Roger hikes 6 miles north, 5 miles east, and then 4 miles north again. To the nearest tenth of a mile, how far, in a straight line, is Roger from his starting point?

_____

13. Where do you find absolute value on the graphing calculator?

_____

14. Evaluate $|x^7-8x+2|$ for x=-5.

_____

15. The table below gives the average high temperature in Fort Drum, NY for each month. Using the data, create a stem-and-leaf plot.

Month	Jan	Feb	Mar	Apr	May	Jun	Jul	Aug	Sep	Oct	Nov	Dec
Temp (F)	27	30	39	52	67	75	79	77	69	56	44	32

16. The list at right gives the regular hunting season totals for black bears killed in the 2001 hunting season in the Adirondack Range. Fill in the table and create a box-and-whisker plot that describes the data.

County	Bears killed
Clinton	11
Essex	42
Franklin	21
Fulton	15
Hamilton	28
Herkimer	10
Jefferson	5
Lewis	21
Oneida	5
St. Lawrence	18
Saratoga	3
Warren	32
Washington	2

Minimum	
1st Quartile	
Median	
3rd Quartile	
Maximum	

17. Find the equation of the line through the points (5,7) and (-15,36).

----

18. Find the equation of the line through the points (6,4) and (13,4).

----

19. Find the equation of the "line of best fit" for the points

(2,5), (1,10), (-2,17), (5,0)

----

20. Find the solution to the following system of equations:

Y=5x-3

Y=-4x+2

----

21. Find the solution(s): $x^2-8x+10=0$

----

22. Find the solution(s): $-2x^2-3x+4=0$

----

23. Draw a circle whose center is (4,2) and has radius 6.

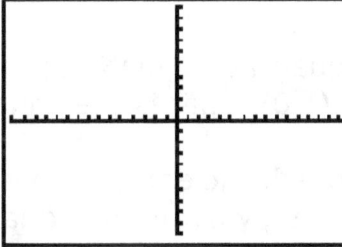

24. Draw a circle whose equation is $(x-3)^2+(y+7)^2=36$

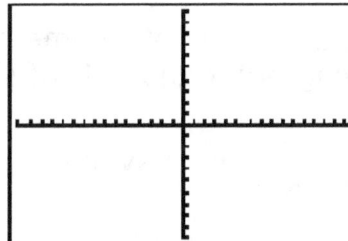

25. Which is an equation of the parabola shown?

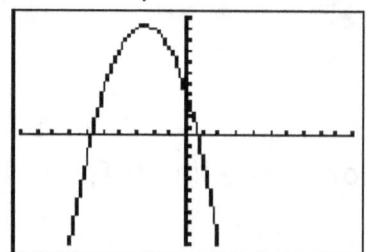

    (1) $x^2+5x+4=0$ (3) $x^2-x+11=0$
    (2) $-x^2-4x+10=0$ (4) $-x^2-5x+4=0$

26. What is the area of a circle with diameter of 10cm? (Leave answer in terms of π.)

                                           ----

27. What is the radius of a circle with circumference 14π in.?

                                           ----

28. What is the total number of points of intersection in the graphs of the equations $x^2+y^2=49$ and $y=7$?

                                           ----

29. Solve :

$$y=2x+4$$
$$y=x+5$$

                                         ----

30. Solve:

$$3x+y=-9$$
$$x+3y=13$$

                                         ----

31. If a quadratic equation has roots that are 4 and –2, what are the factors of the equation?

                                         ----

32. If the graph of a quadratic equation passes through the points (-1,4), (0, -3), and (1,0), what is the equation?

                                         ----

33. If the graph of a quadratic equation passes through the points (-1,-7), (0,3), and (1,9), what is the equation?

                                         ----

34. What are the turning point and axis of symmetry of
$y = 3x^2 + x - 5$?

    TP= _____   axis of symmetry: _____

35. What are the turning point and axis of symmetry of
$y = -x^2 - 5x + 7$?

    TP= _____ axis of symmetry: _____

Review VI

Unless otherwise specified, round answers to the **nearest hundredth.**

1. Add: $11\frac{3}{13}+8\frac{7}{8}$ (express answer as a fraction) _____

2. Subtract: $\frac{17}{18}-\frac{3}{11}$ (express answer as a fraction) _____

3. Multiply: $3\frac{1}{9}\cdot23\frac{4}{7}$ (express answer as a fraction) _____

4. Find the LCM: 13,42,12 _____

5. Find the GCF: 48,96,142 _____

6. Change .82 to a fraction _____

7. Solve for x: 3x-23=89 _____

8. Solve for x: $\frac{x-8}{5}=\frac{7x+2}{3}$ _____

9. Change the unrounded solution in number 8 to a fraction

_____

10. 7 is 30% of what number? _____

11. A wall is supported by a brace 13 feet long as shown in the diagram below. If one end of the brace is placed 5 feet from the base of the wall, how many feet up the wall does the brace reach?

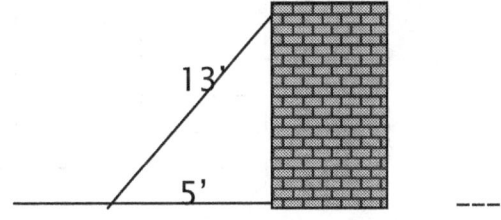

_____

12. What is the value of $x^5-14x^3+5x^2-13$ for x=-3? _____

13. What is the value of $\left|4x^5-30\right|-\left|2x^{-3}+2\right|+12x$ for x=-2? _____

14. The following are scores from the Math A1 final exam:

    77, 42, 89, 67, 100, 75, 80, 88, 90, 95, 70, 82, 69,
    63, 74, 88, 90, 95, 78, 85

    Mean=   _____

    Median=   _____

    Mode=   _____

    Q1=   _____

    Q3=   _____

15. Using the data from #14 complete the frequency table.

Interval	Frequency
40-49	
50-59	
60-69	
70-79	
80-89	
90-100	

16. Using the table above create a frequency histogram which displays the data.

17. Find the roots of the equation $x^2-x-132=0$ _____

18. Find the roots of the equation $x^2-24x+63=0$ _____

19. Change the ZOOM setting to Zsquare. DRAW a circle with a

    center at

    (4,-3) and a radius of 5. Sketch the screen below:

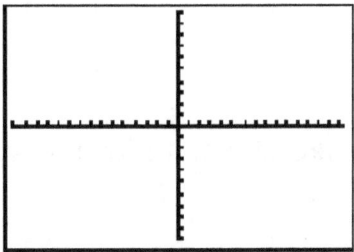

20. What is the center and radius of a circle whose equation is

    $(x-4)^2+(y+1)^2=49$?

    _____ _____

21. Solve the following system of equations:

    $y=(2/3)x-3$
    $y=-2x+5$

    _____

22. Solve the following system of equations:

    $y=2x+4$
    $y=2x-6$

    _____

23. Find the equation of the line through the points (4,-2) and (1,8).

    _____

24. What is the slope of the line through the points (5,0) and

    (5,-10)?

    _____

25. Graph and sketch:
    $y<2x-5$

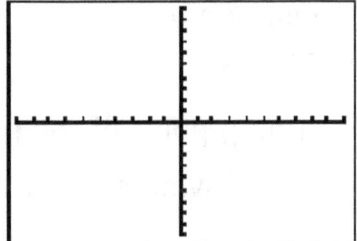

26. Graph and sketch:
$$3x+2y \geq 12$$

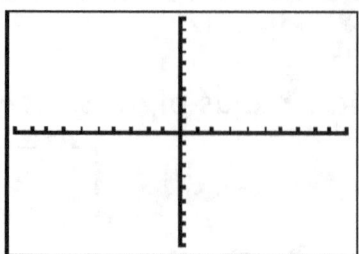

27. Graph and sketch the following system and label the solution set S.

$$y > -6x+9$$
$$y \leq -x^2+5x-2$$

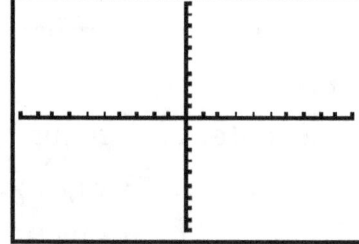

28. A total of 600 tickets were sold for a concert. Twice as many tickets were sold in advance than were sold at the door. If the tickets sold in advance cost $25 each and the tickets sold at the door cost $32 each, how much money was collected for the concert?

----

29. Which of the following are functions?(There may be more than one!)
    a. $x^2-3x+4=y$
    b. $3y^2-6x+2=4$
    c. $x^2+y^2=16$
    d. $x=5$
    e. $y=18$

30. Find the equation of the parabola that passes through points (-1,-15), (0,-3), and (1,7).

----

31. Find the equation of the parabola that passes through points (-1,-1), (0,-3), (1,1).

----

32. Find the turning point of the parabola in #30. --------------

33. Find the axis of symmetry for the parabola in #31.--------------

Review VII

Unless otherwise specified, round answers to the **nearest hundredth**.

1. Subtract (write answer as a mixed number):

    $$3\frac{1}{12} - 7\frac{2}{3}$$ _____

2. Multiply (write answer as a mixed number):

    $$4\frac{2}{5} * 5\frac{3}{7}$$ _____

3. Solve for x: $\dfrac{8x+3}{6} = \dfrac{2x-4}{5}$ _____

4. Write the unrounded answer to number 3 as a fraction:

                                                _____

5. Melissa wants to know the distance across the lake before she
    tries to canoe across. Trail A is marked on her map as 5 miles.
    Trail B is 11 miles and meets Trail A at about a 90 degree angle.
    How far is it across the lake to the nearest tenth of a mile?

    Trail A   Trail B   _____

6. Marvin's mom sees the perfect sweater to buy him for
    Christmas. She pays a total of $44.94. This includes 7% sales
    tax. The tax was applied after her $10.00 coupon that was
    deducted after a 20% sale deduction. What was the original
    price of the sweater?

                                                _____

7. Convert 17.8°C to the nearest tenth of a degree Fahrenheit
    using the formula F°=(9/5)C°+32.

                                                _____

8. Morris found the mass of a sample in chemistry class as 2.6 grams. Mr. Widrick said the mass of the sample was 2.75 grams. What was Morris' percent error?

_____

9. According to the National Oceanic and Atmospheric Administration, the following temperatures represent the average global temperature per decade from 1880 to 1999.

Decade	Temp (F)
1880-89	56.65
1890-99	56.64
1900-09	56.52
1910-19	56.57
1920-29	56.74
1930-39	57.00
1940-49	57.13
1950-59	57.06
1960-69	57.05
1970-79	57.04
1980-89	57.36
1990-99	57.64

Find the equation of the line of best fit for the data.

_____

What would the expected temperature be in the decade 2020-2029?

_____

When would the global temperature reach an average temperature of 60°F?

_____

10. The average surface temperature of the sun has been calculated to be 9941°F. It is theorized that the internal temperature reaches 28,000,000°F. Write each of these temperatures in scientific notation.

_____   _____

Using the correct number of sig figs, write their difference in scientific notation.

_____

11. Based on the answer to #10, what conclusion could be made about the relationship between the surface temperature and the internal temperature of the sun?

----

12. Graph and correctly label the following system of equations:
    y=(1/3)x+5
    y=2x-3

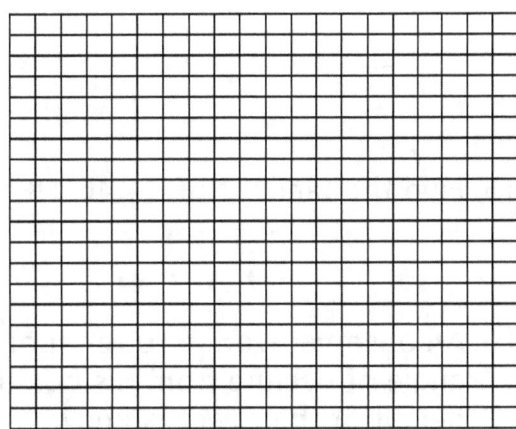

13. Graph and label the following system of inequalities:
    y<-5x+1
    y≥3x+2

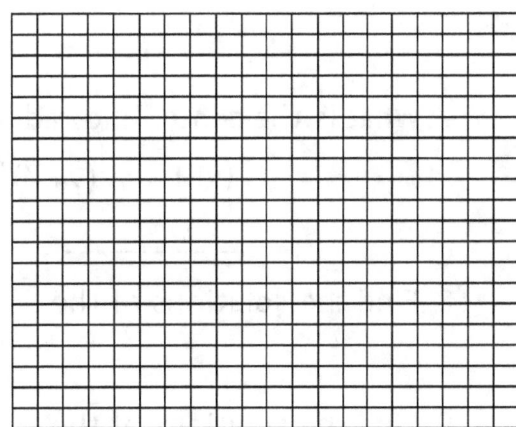

14. Graph and label the following linear/quadratic system:

$$y=x-6$$
$$y=x^2+2x-8$$

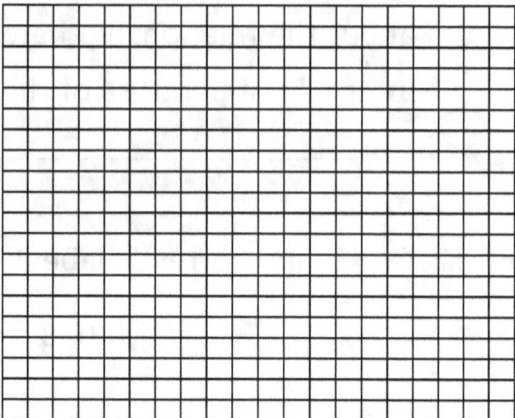

15. Find the tangent of 34°46'18"   _____

16. An observer from a tower spots a forest fire. He notes that the angle of decline from his position to the fire is 1.2° and the height of the tower is known to be 120 feet. When he radios for help how could he most accurately describe the location of the fire?

----

17. What is the center and radius of a circle whose equation is

$$(x-13)^2+(y+7)^2=25?$$

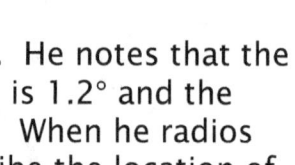

_____   _____

18. Find the solutions of the equation $2x^2-14x-3=0$.

_____   _____

19. Find the solutions of the equation $x^2+3x-5=0$.

_____   _____

20. What is the equation of the line through the points (3,-4) and (11, 5)?

_____

21. What is the slope of the line through the points (5,-1) and (16, -1)? _____

22. What is the slope of the line through the points (5,-1) and (5,17)? _____

23. Simplify: $2.345 \times 10^5 \times 4.57 \times 10^3$ _____

24. Simplify: $5.679 \times 10^{-5} + 2.1 \times 10^{-3}$ _____

25. Simplify: $\dfrac{-2.154 \times 10^9}{1.527 \times 10^{-1}}$ _____

For #26-28 assume that triangle ABC is a right triangle.

26. Find the length of AB to the nearest foot:

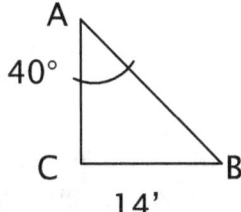

_____

27. Find the measure of angle B to the nearest tenth of a degree.

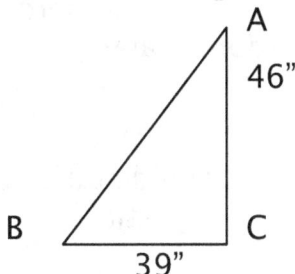

28. Find the measure of BC to the nearest meter. _____

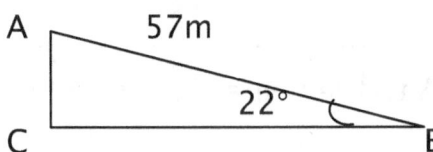

_____

29. Draw and label a diagram of the path of an airplane climbing at an angle of 11° with the ground. Find, to the *nearest foot*, the ground distance the airplane has traveled when it has attained an altitude of 400 feet.

_____

30. A ship on the ocean surface detects a sunken ship on the ocean floor at an angle of depression of 50°. The distance between the ship on the surface and the sunken ship on the ocean floor is 200 meters. If the ocean floor is level in this area, how far above the ocean floor, to the *nearest meter*, is the ship on the surface?

----

31. Utica, NY and Philadelphia, PA are at approximately the same longitude. If Utica's latitude is 43°6'3"N and Philadelphia's latitude is 39°57'8"N, how far is it between the two cities?

    a. In degrees/minutes/seconds: _____

    b. In miles: (one degree latitude = _____ miles)

----

32. Rhonda weighs herself on her home bathroom scale that says she weighs 129 lbs. At the doctor's office a few minutes later the scale reads 134 lbs. She is assured that the doctor's scale has been correctly calibrated. What is the percent error for her weight as found with her bathroom scale?

----

33. Why is it harder to find the horizontal distance between two cities that the vertical distance using latitude and longitude?

----

34. Find the length of CA to the nearest *centimeter*.

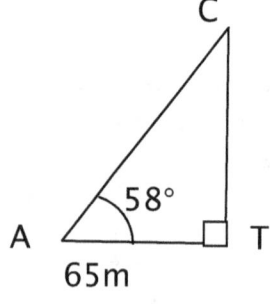

Unless otherwise specified, round answers to the nearest hundredth.

1. Subtract: $2\frac{1}{7} - 5\frac{4}{5}$ (Express answer as a mixed number.)

   _____

2. Solve for x: 4x-3=2x+7   _____

3. If a construction worker wants a 20-foot ladder to rest exactly 16 feet up the side of a building, how far from the building must the base of the ladder be placed?

   _____

4. Simplify 8!   _____

5. Simplify $_7C_4$ _____

6. Simplify $_7P_4$ _____

7. Explain why $_7C_4$ is less than $_7P_4$.

   _____

   _____

   _____

8. A DJ has a list of 10 requested songs. How many choices does he have for the order in which to play them?

   _____

9. It's getting late and the DJ now has a list of 20 requests, but only time to play 5. How many combinations does he have to choose from for his last 5 songs of the night?

   _____

10.   An airplane crashed after its pitch (angle of inclination) changed from a "normal" 7° to 52° by the time it reached 1200'. What would the approximate difference in ground distance traveled be between a plane taking off and remaining at the normal pitch and this doomed plane at its 52° pitch when the planes reached 1200'?

    ----

11.   Write the equation of the line that passes through the points (7,-2) and (15,3).

    ----

12.   What is the slope of the line that passes through the points (6,2) and   (-15,12)?

    ----

13.   What is the slope of the line that passes through the points (4,9) and (4,16)?

    ----

14.   If x=-5, what is the value of $|2x^3-10|$?

    ----

15.   Find the solutions for the quadratic equation $0=x^2-4x-12$

    ----

16.   Graph $y=x^2-6x+5$ and $y=3x+4$ on the axes below. Label appropriately.

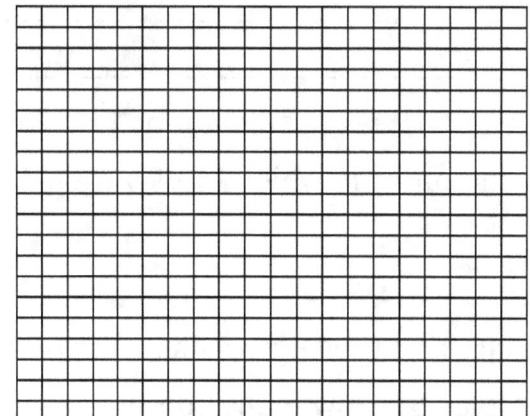

17.   If an angle measures 46.79 degrees, what does this same angle measure   in degrees, minutes, and seconds?

    ----

18.   The electoral votes for the last ten presidential elections
     are listed below:

1964	Lyndon Johnson	486
1968	Richard M. Nixon	301
1972	Richard M. Nixon	520
1976	Jimmy Carter	297
1980	Ronald Reagan	489
1984	Ronald Reagan	525
1988	George H. W. Bush	426
1992	Bill Clinton	370
1996	Bill Clinton	379
2000	George W. Bush	271

Find the mean number of electoral votes received.

Create a box-and-whisker plot to display the data.

<-->

19.   Draw the circle described by the equation

$(x-2)^2+(y+6)^2=16$.

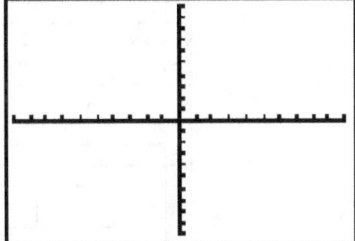

20.   If you have 4 quarters, 5 dimes, 3 nickels, and 6 pennies
     in your pocket and 4 coins fell out when you pulled your pen
     from your pocket, what is the probability that you lost at least
     $.75?

21. Write the equation of the parabola that passes through the points (-2,7), (0,1), and (2,-21).

_____

22. Write the equation of the parabola that passes through the points (-1,26), (0,3), and (1,-10).

_____

23. Find the turning point and axis of symmetry for the equation in **#21**.

Turning point: _____
Axis of Symmetry: _____

24. Find the turning point and axis of symmetry for the equation found in **#22**.

Turning point: _____

Axis of Symmetry: _____

25. If the solutions of a quadratic equation are –2 and 14, what are the factors?

_____

26. If the roots of a quadratic equation are 5 and –7, what are the factors?

_____

27. Sketch:
   $y<x-1$

28. Sketch:
   $y\geq2x+1$

29. Show the solution set:

$y < -x + 7$
$y \geq 2x + 1$

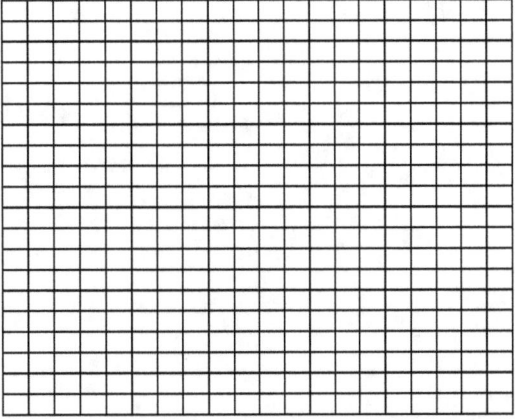

Show Shading!

www.ingramcontent.com/pod-product-compliance
Lightning Source LLC
Chambersburg PA
CBHW081109170526
45165CB00008B/2390